STATUTORY INSTRUMENTS

1999 No. 2864

ROAD TRAFFIC

The Motor Vehicles (Driving Licences) Regulations 1999

Made - - - - -	*19th October 1999*
Laid before Parliament	*21st October 1999*
Coming into force - -	*12th November 1999*

ARRANGEMENT OF REGULATIONS

PART I

PRELIMINARY

PART II

LICENCES

Categories of entitlement

Minimum ages for holding and obtaining licences

Applications for licences

[DOT 8837]

1

PART III

TESTS OF COMPETENCE TO DRIVE

PART IV

GOODS AND PASSENGER-CARRYING VEHICLES

General

Persons under the age of 21

Drivers' conduct

PART V

APPROVED TRAINING COURSES FOR RIDERS OF MOTOR BICYCLES AND MOPEDS

Approved training courses

Instructors

Miscellaneous

PART VI

DISABILITIES

70. Licence groups
71. Disabilities prescribed in respect of Group 1 and 2 licences
72. Disabilities prescribed in respect of Group 1 licences
73. Disabilities prescribed in respect of Group 2 licences
74. Disabilities requiring medical investigation: High Risk Offenders
75. Examination by an officer of the Secretary of State

PART VII

SUPPLEMENTARY

Transitional provisions

76. Effect of change in classification of vehicles for licensing purposes
77. Saving in respect of entitlement to Group M
78. Saving in respect of entitlement to former category N
79. Saving in respect of entitlement to drive mobile project vehicles

Miscellaneous

80. Persons who become resident in Great Britain
81. Service personnel
82. Northern Ireland licences
83. Statement of date of birth

SCHEDULES

1. Regulations revoked
2. Categories and sub-categories of vehicle for licensing purposes
3. Licence fees
4. Distinguishing marks to be displayed on a vehicle being driven under a provisional licence
5. Fees for practical and unitary tests
6. Evidence of identity of test candidates
7. Specified matters for theory test
8. Specified requirements for practical or unitary test
9. Upgraded entitlements on passing second test etc
10. Forms of certificate and statement of theory test result
11. Forms of certificate and statement of practical and unitary test result
12. Elements of an approved training course
13. Approved motor bicycle training courses: forms of certificate

The Secretary of State for the Environment, Transport and the Regions, in exercise of the powers conferred by the following sections of the Road Traffic Act 1988(a), namely

---

(a) 1988 c. 52. Relevant amendments are referred to where appropriate below. Amending statutes and instruments are referred to in the footnotes to these Regulations in the following manner: "the 1989 Act" means the Road Traffic (Driver Licensing and Information Systems) Act 1989 (c. 22); "the 1990 Regulations" means the Driving Licences (Community Driving Licence) Regulations 1990 (S.I. 1990/144), "the 1991 Act" means the Road Traffic Act 1991 (c. 40), "the 1996 Regulations" means the Driving Licences (Community Driving Licence) Regulations 1996 (S.I. 1996/1974) and "the 1998 Regulations" means the Driving Licences (Community Driving Licence) Regulations 1998 (S.I. 1998/1420).

sections 88(5) and (6)(**a**), 89(1A), (2A), (3), (4), (5), (5A), (6), (7), (9) and (10)(**b**), 89A(3) and (5)(**c**), 91, 92(2) and (4)(**d**), 94(4) and (5)(**e**), 97(1), (1A), (3), (3A), (3B) and (4)(**f**), 98(2) and (4)(**g**), 99(1) and (1A)(**h**), 99A(3), (4) and (6)(**i**), 101(2) and (3), 105(1), (2), (3) and (4)(**j**), 108(1)(**k**), 114(1)(**l**), 115(1) and (3)(**m**), 115A(1)(**n**), 117(2A)(**o**), 118(4), 120, 121(**p**), 122(**q**), 164(2)(**r**), 183(6)(**s**) and 192(1)(**t**), after consulting with representative organisations in accordance with section 195(2) of the Road Traffic Act 1988 and, in the case of regulations 14, 30 and 35, with the approval of the Treasury(**u**), hereby makes the following Regulations:

PART I

PRELIMINARY

Citation and commencement

1. These Regulations may be cited as the Motor Vehicles (Driving Licences) Regulations 1999 and shall come into force on 12th November 1999.

Revocation and saving

2.—(1) The regulations specified in Schedule 1 are hereby revoked.

(2) Subject to otherwise herein provided, and without prejudice to the operation of sections 16 and 17 of the Interpretation Act 1978(**v**), the revocation of those regulations shall not affect the validity of any application or appointment made, notice or approval given, licence, certificate or other document granted or issued or other thing done thereunder and any reference in such application, appointment, notice, approval, licence, certificate or other document or thing to a provision of any regulation hereby revoked, whether specifically or by means of a general description, shall, unless the context otherwise requires, be construed as a reference to the corresponding provision of these Regulations.

(**a**) Subsection (6) was amended by the 1990 Regulations.
(**b**) Subsection (1A) was inserted by the 1996 Regulations, subsections (2A) and (5A) were inserted by the 1989 Act, section 6; subsection (3) was amended by the 1991 Act, Schedule 4, paragraph 63; subsection (4) was amended by the 1989 Act, Schedule 3, paragraph 8 and by the 1996 Regulations; subsection (7) was amended and subsections (9) and (10) were substituted by the 1989 Act, Schedule 3, paragraph 8 and subsections (7), (9) and (10) were amended by the 1996 Regulations. Subsection (4)(b) is to be read with the Department of Transport (Fees) Order 1988 (S.I. 1988/643), as amended by S.I. 1991/811, 1993/1601, 1995/1684 and 1996/1929, the relevant entries being items 5, 5A and 6 in Schedule 1, Table III.
(**c**) Section 89A was inserted by the 1989 Act, section 4(4).
(**d**) Subsection (2) was amended by the 1996 Regulations.
(**e**) Subsections (4) and (5) were amended by the 1989 Act, section 5(7) and (8).
(**f**) Subsections (1) and (3) were amended by the 1989 Act, section 6(2) and Schedule 6 and by the 1996 and 1998 Regulations, subsection (1A) was inserted by the 1998 Regulations; subsections (3A) and (3B) were inserted by the 1989 Act, section 6(2) and subsection (3A) was amended by the 1996 and 1998 Regulations; subsection (4) was amended by the 1996 and 1998 Regulations.
(**g**) Subsections (2) and (4) were amended by the 1989 Act, Schedule 3, paragraph 11. Subsection (2) was amended by the 1990 Regulations.
(**h**) Subsection (1) was amended by Schedule 3, paragraph 12, of the 1989 Act; subsection (1A) was inserted by section 2(2) of that Act.
(**i**) Section 99A was inserted by the 1996 Regulations; subsection (4) was amended by the 1998 Regulations.
(**j**) Subsection (2) was amended by the 1989 Act, Schedule 3, paragraph 14 and by the 1990, 1996 and 1998 Regulations; subsection (4) was amended by the 1996 Regulations.
(**k**) See the definitions of "prescribed" and "regulations".
(**l**) Sections 114, 115, 116 and 117 to 122 (Part IV of the Act) were substituted by the 1989 Act, section 2(1) and Schedule 2. Section 114(1) was amended by the 1996 Regulations.
(**m**) Subsection (3) was amended by the 1991 Act, Schedule 4, paragraph 64 and the 1996 Regulations.
(**n**) Section 115A was inserted by the 1996 Regulations.
(**o**) Subsection (2A) was inserted by the 1991 Act, Schedule 4, paragraph 65.
(**p**) See the definition of "prescribed".
(**q**) Amended by the 1990 Regulations.
(**r**) Amended by the 1991 Act, Schedule 4, paragraph 68.
(**s**) Inserted by the 1989 Act, Schedule 3, paragraph 23.
(**t**) See the definition of "prescribed".
(**u**) See section 105(4) of the Road Traffic Act 1988.
(**v**) 1978 c. 30.

Interpretation

3.—(1) In these Regulations, unless the context otherwise requires, the following expressions have the following meanings—

"1981 Act" means the Public Passenger Vehicles Act 1981(**a**);

"1985 Act" means the Transport Act 1985(**b**);

"ambulance" means a motor vehicle which—

(a) is constructed or adapted for, and used for no other purpose than, the carriage of sick, injured or disabled people to or from welfare centres or places where medical or dental treatment is given, and

(b) is readily identifiable as such a vehicle by being marked "Ambulance" on both sides;

"appropriate driving test" and "extended driving test" have the same meanings respectively as in section 36 of the Offenders Act(**c**);

"Construction and Use Regulations" means the Road Vehicles (Construction and Use) Regulations 1986(**d**);

"controlled by a pedestrian" in relation to a vehicle means that the vehicle either—

(a) is constructed or adapted for use under such control; or

(b) is constructed or adapted for use either under such control or under the control of a person carried on it but is not for the time being in use under, or proceeding under, the control of a person carried on it;

"dual purpose vehicle" means a motor vehicle which is constructed or adapted both to carry or haul goods and to carry more than eight persons in addition to the driver;

"exempted goods vehicle" and "exempted military vehicle" have the meanings respectively given in regulation 51;

"extended driving test" means a test of a kind prescribed by regulation 41;

"full", in relation to a licence of any nature, means a licence granted otherwise than as a provisional licence;

"Group 1 licence" and "Group 2 licence" have the meanings respectively given in regulation 70;

"incomplete large vehicle" means—

(a) an incomplete motor vehicle, typically consisting of a chassis and a complete or incomplete cab, which is capable of becoming, on the completion of its construction, a medium-sized or large goods vehicle or a passenger-carrying vehicle, or

(b) a vehicle which would be an articulated goods vehicle but for the absence of a fifth-wheel coupling,

and which is not drawing a trailer;

"large motor bicycle" means—

(a) in the case of a motor bicycle without a side-car, a bicycle the engine of which has a maximum net power output exceeding 25 kilowatts or which has a power to weight ratio exceeding 0.16 kilowatts per kilogram, or

(b) in the case of a motor bicycle and side-car combination, a combination having a power to weight ratio exceeding 0.16 kilowatts per kilogram;

"LGV trainee driver's licence" has the meaning given in regulation 54;

"maximum authorised mass"—

(a) in relation to a goods vehicle, has the same meaning as "permissible maximum weight" in section 108(1) of the Traffic Act,

(b) in relation to an incomplete large vehicle, means its working weight, and

(c) in relation to any other motor vehicle or trailer, has the same meaning as "maximum gross weight" in regulation 3(2) of the Construction and Use Regulations;

"maximum speed" means the speed which the vehicle is incapable, by reason of its construction, of exceeding on the level under its own power when fully laden;

"maximum net power output" has the same meaning as in section 97 of the Traffic Act;

(**a**) 1981 c. 14.
(**b**) 1985 c. 67.
(**c**) Section 36 was substituted by the 1991 Act, section 32.
(**d**) S.I. 1986/1078. The relevant amending instruments are S.I. 1987/676, 1990/1981 and 1994/329.

"mobile project vehicle" means a vehicle which has a maximum authorised mass exceeding 3.5 tonnes, is constructed or adapted to carry not more than eight persons in addition to the driver and carries principally goods or burden consisting of—

(a) play or educational equipment and articles required in connection with the use of such equipment, or

(b) articles required for the purposes of display or of an exhibition,

and the primary purpose of which is use as a recreational, educational or instructional facility when stationary;

"Northern Ireland test" means a test of competence to drive conducted under the law of Northern Ireland;

"Offenders Act" means the Road Traffic Offenders Act 1988(a);

"passenger-carrying vehicle recovery vehicle" means a vehicle (other than an articulated goods vehicle combination as defined in section 108(1) of the Traffic Act) which—

(a) has an unladen weight not exceeding 10.2 tonnes,

(b) is being operated by the holder of a PSV operator's licence, and

(c) is being used for the purpose of—

(i) proceeding to, or returning from, a place where assistance is to be, or has been, given to a damaged or disabled passenger-carrying vehicle; or

(ii) giving assistance to or moving a disabled passenger-carrying vehicle or moving a damaged vehicle;

"penalty points" means penalty points attributed to an offence under section 28 of the Offenders Act;

"power to weight ratio", in relation to a motor bicycle, means the ratio of the maximum net power output of the engine of the vehicle to its weight (including the weight of any side-car) with—

(a) a full supply of fuel in the tank,

(b) an adequate supply of other liquids needed for its propulsion, and

(c) no load other than its normal equipment, including loose tools;

"practical test" means a practical test of driving skills and behaviour or, where a test is by virtue of these Regulations required to be conducted in two parts, the part of it which consists of that test and includes such a test conducted as part of an extended driving test;

"propelled by electrical power", in relation to a motor vehicle, means deriving motive power solely from an electrical storage battery carried on the vehicle and having no connection to any other source of power when the vehicle is in motion;

"PSV operator's licence" has the meaning given by section 82(1) of the 1981 Act;

"standard access period" has the meaning given by regulation 22;

"standard motor bicycle" means a motor bicycle which is not a large motor bicycle;

"test" means any test of competence to drive conducted pursuant to section 89 of the Traffic Act including an extended driving test;

"test pass certificate" means a certificate in the form specified in regulation 48(1)(a);

"theory test" means, where a test is by virtue of these Regulations to be conducted in two parts, the part that consists of the theoretical test and includes such a test conducted as part of an extended driving test;

"theory test pass certificate" means a certificate in the form specified in regulation 47(2)(a);

"Traffic Act" means the Road Traffic Act 1988;

"traffic commissioner" means, in relation to an applicant for or the holder of a licence, the traffic commissioner in whose area the applicant or holder resides;

"unitary test" means a test which, by virtue of these Regulations, is to consist of a single test of both practical driving skills and behaviour and knowledge of the Highway Code and other matters and includes such a test conducted as an extended driving test;

"unladen weight" has the same meaning as in regulation 3(2) of the Construction and Use Regulations and, in the case of a road roller, includes the weight of any object for the time being attached to the vehicle, being an object specially designed to be so attached for the purpose of temporarily increasing the vehicle's weight;

"vehicle with automatic transmission" means a class of vehicle in which either—

(a) 1988 c. 53.

(a) the driver is not provided with any means whereby he may vary the gear ratio between the engine and the road wheels independently of the accelerator and the brakes, or

(b) he is provided with such means but they do not include a clutch pedal or lever which he may operate manually,

(and accordingly a vehicle with manual transmission is any other class of vehicle);

"working weight" means the weight of a vehicle in working condition on a road but exclusive of the weight of any liquid coolant and fuel used for its propulsion.

(2) In these Regulations, unless the context otherwise requires—

(a) a reference to a licence being in force is a reference to it being in force in accordance with section 99 of the Traffic Act, save that for the purpose of these Regulations a licence shall remain in force notwithstanding that it is—

(i) surrendered to the Secretary of State or is revoked otherwise than by notice under section 93(1) or (2) of the Traffic Act (revocation because of disability or prospective disability), or

(ii) treated as revoked by virtue of section 37(1) of the Offenders Act, and

(b) a reference to the expiry of a licence is a reference to the time at which it ceases to be so in force (and "expired" shall be construed accordingly).

(3) Except where otherwise expressly provided, any reference in these Regulations to a numbered regulation or Schedule is a reference to the regulation or Schedule bearing that number in these Regulations, and any reference to a numbered paragraph (otherwise than as part of a reference to a numbered regulation) is a reference to the paragraph bearing that number in the regulation or Schedule in which the reference occurs.

(4) Where a statement or certificate (but not a distinguishing mark specified in regulation 16) is required under these Regulations to be in a form prescribed herein, the reference is to a certificate or statement in that form (or as nearly in that form as circumstances permit), adapted to the circumstances of the case and duly completed and signed where required.

(5) For the purposes of section 97(3)(d) of the Traffic Act and these Regulations the date of first use of a motor bicycle means—

(a) except in a case to which paragraph (b) applies, the date on which it was first registered under the Roads Act 1920, the Vehicles (Excise) Act 1949(**a**), the Vehicles (Excise) Act 1962(**b**) or the Vehicles (Excise) Act 1971(**c**);

(b) in the case of a motor bicycle which was used in any of the following circumstances before the date on which it was first registered, namely:—

(i) where the bicycle was used under a trade licence as defined in section 16 of the Vehicles (Excise) Act 1971, otherwise than for the purposes of demonstration or testing or of being delivered from premises of the manufacturer by whom it was made, or of a distributor of vehicles or dealer in vehicles to premises of a distributor of vehicles, dealer in vehicles or purchaser thereof, or to premises of a person obtaining possession thereof under a hiring agreement or hire purchase agreement,

(ii) where the bicycle belonged to the Crown and is or was used or appropriated for use for naval, military or air force purposes,

(iii) where the bicycle belonged to a visiting force or a headquarters or defence organisation to which the Visiting Forces and International Headquarters (Application of Law) Order 1965(**d**) applied,

(iv) where the bicycle had been used on roads outside Great Britain and was imported into Great Britain, or

(v) where the bicycle had been used otherwise than on roads after being sold or supplied by retail and before being registered,

the date of manufacture of the bicycle.

(6) In paragraph (5)(b)(v) "sold or supplied by retail" means sold or supplied otherwise than to a person acquiring solely for the purpose of re-sale or re-supply for a valuable consideration.

(**a**) 1949 c. 89.
(**b**) 1962 c. 13.
(**c**) 1971 c. 10.
(**d**) S.I. 1965/1536.

PART II

LICENCES

Categories of entitlement

Classification of vehicles

4.—(1) Subject to regulations 5 and 78, the Secretary of State shall grant licences authorising the driving of motor vehicles in accordance with the categories and sub-categories specified in column (1) and defined in column (2) of Schedule 2 and those categories and sub-categories are designated as groups for the purposes of section 89(1)(b) of the Traffic Act.

(2) In these Regulations, expressions relating to vehicle categories have the following meanings—

    (a) any reference to a category or sub-category identified by letter, number or word or by any combination of letters, numbers and words is a reference to the category or sub-category defined in column (2) of Schedule 2 opposite that letter or combination in column (1) of the Schedule,

    (b) "sub-category" means, in relation to category A, B, C, C + E, D or D + E, a class of vehicles comprising part of the category and identified as a sub-category thereof in column (2) of Schedule 2, and

    (c) unless the context otherwise requires, a reference to a category includes a reference to sub-categories of that category.

Classes for which licences may be granted

5.—(1) A licence authorising the driving of motor vehicles of a class included in a category or sub-category shown in Part 1 of Schedule 2 may be granted to a person who is entitled thereto by virtue of—

    (a) holding or having held a full licence, a full Northern Ireland licence, full British external licence, full British Forces licence, exchangeable licence or Community licence authorising the driving of vehicles of that class, or

    (b) having passed a test for a licence authorising the driving of motor vehicles of that class or a Northern Ireland or Gibraltar test corresponding to such a test.

(2) A licence authorising the driving of motor vehicles of a class included in any category or sub-category shown in Part 2 of Schedule 2 may not be granted to a person unless, at a time before 1st January 1997—

    (a) in the case of a person applying for a full licence,—

        (i) he held a full licence authorising the driving of motor vehicles of that class or a class which by virtue of these Regulations corresponds to a class included in that category or sub-category, or

        (ii) he passed a test which at the time it was passed authorised the driving of motor vehicles of such a class or a Northern Ireland test corresponding to such a test;

    (b) in the case of a person applying for a provisional licence, he held a provisional licence authorising the driving of vehicles of that class or a class which by virtue of these Regulations corresponds to a class included in that category or sub-category.

(3) A licence authorising the driving of motor vehicles included in sub-category B1 (invalid carriages), which are specified in Part 3 of Schedule 2, may not be granted to a person unless, at a time before 12th November 1999—

    (a) in the case of a person applying for a full licence, he held a full licence authorising the driving of motor vehicles included in sub-category B1 (invalid carriages) or a class of motor vehicles which by virtue of these Regulations corresponds to vehicles included in that sub-category, or

    (b) in the case of a person applying for a provisional licence, he held a provisional licence authorising the driving of motor vehicles included in sub-category B1 (invalid carriages) or a class of motor vehicles which by virtue of these Regulations corresponds to vehicles included in that sub-category.

Competence to drive classes of vehicle: general

6.—(1) Where a person holds, or has held, a relevant full licence authorising him to drive vehicles included in any category or, as the case may be, sub-category he is deemed competent to drive—

    (a) vehicles of all classes included in that category or sub-category unless by that licence he is or was authorised to drive—

        (i) only motor vehicles of a specified class within that category or sub-category, in which case he shall be deemed competent to drive only vehicles of that class, or

        (ii) only motor vehicles adapted on account of a disability, in which case he shall be deemed competent to drive only such classes of vehicle included in that category or sub-category as are so adapted (and for the purposes of this paragraph, a motor bicycle with a side-car may be treated in an appropriate case as a motor vehicle adapted on account of a disability),

and

    (b) all classes of vehicle included in any other category or sub-category which is specified in column (3) of Schedule 2 as an additional category or sub-category in relation to that category or sub-category unless by that licence he is or was authorised to drive—

        (i) only motor vehicles having automatic transmission, in which case he shall, subject to paragraph (2), be deemed competent to drive only such classes of motor vehicle included in the additional category or sub-category as have automatic transmission, or

        (ii) only motor vehicles adapted on account of a disability, in which case he shall be deemed competent to drive only such classes of vehicle included in the additional category or sub-category as are so adapted.

(2) Where the additional category is F, K or P, paragraph (1)(b)(i) shall not apply.

(3) In this regulation and regulations 7 and 8, "relevant full licence" means a full licence granted under Part III of the Traffic Act, a full Northern Ireland licence or a Community licence.

Competence to drive classes of vehicle: special cases

7.—(1) A person who has held, for a period of at least two years, a relevant full licence authorising the driving of vehicles included in category C, other than vehicles included in sub-category C1, may also drive a motor vehicle of a class included in category D which is—

    (a) damaged or defective and being driven to a place of repair or being road tested following repair, and

    (b) is not used for the carriage of any person who is not connected with its repair or road testing,

unless by that licence he is authorised to drive only vehicles having automatic transmission, in which case he shall be deemed competent to drive only such of the vehicles mentioned in sub-paragraphs (a) and (b) as have automatic transmission.

(2) A person who holds a relevant full licence authorising the driving of vehicles included in category D, other than vehicles included in sub-category D1 or D1 (not for hire or reward), may drive a passenger-carrying vehicle recovery vehicle unless by that licence he is authorised to drive only vehicles having automatic transmission, in which case he shall be deemed competent to drive only passenger-carrying vehicle recovery vehicles having automatic transmission.

(3) A person may drive an incomplete large vehicle—

    (a) having a working weight exceeding 3.5 tonnes but not exceeding 7.5 tonnes if he holds a relevant full licence authorising the driving of vehicles in sub-category C1, or

    (b) having a working weight exceeding 7.5 tonnes if he holds a relevant full licence authorising the driving of vehicles in category C, other than vehicles in sub-category C1,

unless by that licence he is authorised to drive only motor vehicles having automatic transmission, in which case he shall be deemed competent to drive only incomplete large vehicles of the appropriate weight specified in paragraph (a) or (b) which have automatic transmission.

(4) A person who holds a relevant full licence authorising the driving of vehicles included in category B, other than vehicles in sub-categories B1 and B1 (invalid carriages), may drive—

    (a) an exempted goods vehicle other than—

        (i) a passenger-carrying vehicle recovery vehicle, or

        (ii) a mobile project vehicle,

    (b) an exempted military vehicle, and

    (c) a passenger-carrying vehicle in respect of which the conditions specified in regulation 50(2) or (3) are satisfied,

unless by that licence he is authorised to drive only motor vehicles having automatic transmission, in which case he shall be deemed competent to drive only such of the vehicles mentioned in sub-paragraphs (a), (b) and (c) as have automatic transmission.

(5) A person who—

    (a) holds a relevant full licence authorising the driving of vehicles of a class included in category B, other than vehicles in sub-categories B1 or B1 (invalid carriages),

    (b) has held that licence for an aggregate period of not less than 2 years, and

    (c) is aged 21 or over,

may drive a mobile project vehicle on behalf of a non-commercial body—

        (i) to or from the place where the equipment it carries is to be, or has been, used, or the display or exhibition is to be, or has been, mounted, or

        (ii) to or from the place where a mechanical defect in the vehicle is to be, or has been, remedied, or

        (iii) in such circumstances that by virtue of paragaph 22 of Schedule 2 to the Vehicle Excise and Registration Act 1994(a) the vehicle is not chargeable with duty in respect of its use on public roads,

unless by that licence he is authorised to drive only vehicles having automatic transmission, in which case he shall be deemed competent to drive only mobile project vehicles having automatic transmission.

(6) A person who—

    (a) holds a relevant full licence authorising the driving of vehicles of a class included in category B, other than vehicles in sub-categories B1 or B1 (invalid carriages),

    (b) has held that licence for an aggregate period of not less than 2 years,

    (c) is aged 21 or over,

    (d) if he is aged 70 or over, is not suffering from a relevant disability in respect of which the Secretary of State would be bound to refuse to grant him a Group 2 licence, and

    (e) receives no consideration for so doing, other than out-of-pocket expenses,

may drive, on behalf of a non-commercial body for social purposes but not for hire or reward, a vehicle of a class included in sub-category D1 which has no trailer attached and has a maximum authorised mass—

        (i) not exceeding 3.5 tonnes, excluding any part of that weight which is attributable to specialised equipment intended for the carriage of disabled passengers, and

        (ii) not exceeding 4.25 tonnes otherwise,

unless such a person is by that licence authorised to drive only vehicles having automatic transmission, in which case he shall be deemed competent to drive only such vehicles in sub-category D1 as conform to the above specification and have automatic transmission.

(7) A person who holds a relevant full licence authorising the driving of vehicles of a class included in category B, other than vehicles in sub-categories B1 or B1 (invalid carriages), may drive a vehicle of a class included in category B + E where—

    (a) the trailer consists of a vehicle which is damaged or defective and is likely to represent a road safety hazard or obstruction to other road users,

    (b) the vehicle is driven only so far as is reasonably necessary in the circumstances to remove the hazard or obstruction, and

---

(a) 1994 c. 22.

(c) he receives no consideration for driving the vehicle,

unless by that licence he is authorised to drive only motor vehicles having automatic transmission, in which case he shall be deemed competent to drive, in the circumstances mentioned above, only vehicles included in category B + E having automatic transmission.

Competence to drive classes of vehicle: dual purpose vehicles

8.—(1) Subject to paragraph (2), a person who is a member of the armed forces of the Crown may drive a dual purpose vehicle when it is being used to carry passengers for naval, military or air force purposes—

(a) where the vehicle has a maximum authorised mass not exceeding 3.5 tonnes, if he holds a relevant full licence authorising the driving of vehicles included in category B other than vehicles in sub-categories B1 or B1 (invalid carriages),

(b) where the vehicle has a maximum authorised mass exceeding 3.5 tonnes but not exceeding 7.5 tonnes, if he holds a relevant full licence authorising the driving of vehicles included in sub-category C1,

(c) in any other case, if he holds a relevant full licence authorising the driving of vehicles included in category C other than vehicles in sub-category C1.

(2) Where the person is authorised by his licence to drive only motor vehicles included in the relevant category or sub-category having automatic transmission, he may drive only dual purpose vehicles having automatic transmission.

Minimum ages for holding or obtaining licences

Minimum ages for holding or obtaining licences

9.—(1) Subsection (1) of section 101 of the Traffic Act shall have effect as if for the classes of vehicle and the ages specified in the Table in that subsection there were substituted classes of vehicle and ages in accordance with the following provisions of this regulation.

(2) In item 3 (motor bicycles), the age of 21 is substituted for the age of 17 in a case where the motor bicycle is a large motor bicycle except in the following cases, namely—

(a) a case where a person has passed a test on or after 1st January 1997 for a licence authorising the driving of a motor vehicle of a class included in category A, other than sub-category A1, and the standard access period has elapsed,

(b) a case where the large motor bicycle—

(i) is owned or operated by the Secretary of State for Defence, or

(ii) is being driven by a person for the time being subject to the orders of a member of the armed forces of the Crown

and is being used for naval, military or air force purposes, and

(c) a case where a person holds a licence authorising the driving of a large motor bicycle by virtue of having passed a test before 1st January 1997.

(3) In item 4 (agricultural and forestry tractors), in the case of an agricultural or forestry tractor which—

(a) is so constructed that the whole of its weight is transmitted to the road surface by means of wheels,

(b) has an overall width not exceeding 2.45 metres, and

(c) is driven either—

(i) without a trailer attached to it, or

(ii) with a trailer which has an overall width not exceeding 2.45 metres and is either a two-wheeled or close-coupled four-wheeled trailer,

the age of 16 is substituted for the age of 17 in the case of a person who has passed a test prescribed in respect of category F, or is proceeding to, taking or returning from, such a test.

(4) In item 5 (small vehicles), the age of 16 is substituted for the age of 17 in the case of a small vehicle driven without a trailer attached where the driver of the vehicle is a person in respect of whom an award of the higher rate component of the disability living allowance made in pursuance of section 73 of the Social Security Contributions and Benefits Act 1992(**a**) (whether before or after his 16th birthday) is still in force.

(5) In item 6 (medium-sized goods vehicles), the age of 21 is substituted for the age of 18 in the case of a vehicle drawing a trailer where the maximum authorised mass of the combination exceeds 7.5 tonnes.

(6) In item 7 (other vehicles, including large goods and passenger-carrying vehicles), the age of 18 is substituted for the age of 21 in the case of a person driving a vehicle of a class included in sub-category D1 which is an ambulance and which is owned or operated by—

    (a) a health service body (as defined in section 60(7) of the National Health Service and Community Care Act 1990(**b**)), or

    (b) a National Health Service Trust established under Part I of that Act or under the National Health Service (Scotland) Act 1978(**c**).

(7) In item 7, the age of 18 is substituted for the age of 21 in the case of a motor vehicle and trailer combination which is in sub-category C1 + E and the maximum authorised mass of the combination does not exceed 7.5 tonnes.

(8) In item 7, the age of 18 is substituted for the age of 21 in the case of a person who is registered as an employee of a registered employer in accordance with the Training Scheme, where he is driving a vehicle which is—

    (a) of a class to which his training agreement applies, and

    (b) owned or operated by his employer or by a registered LGV driver training establishment.

(9) In item 7, the age of 18 is substituted for the age of 21 in relation to a passenger-carrying vehicle—

    (a) in the case of a person who holds a provisional licence, and

    (b) in the case of a person who holds a full passenger-carrying vehicle driver's licence, where he is driving a vehicle which is operated under a PSV operator's licence, a permit granted under section 19 of the 1985 Act or a community bus permit granted under section 22 of that Act and he is either—

        (i) not engaged in the carriage of passengers, or

        (ii) engaged in the carriage of passengers on a regular service over a route which does not exceed 50 kilometres, or

        (iii) is driving a vehicle of a class included in sub-category D1.

(10) In items 6 and 7, the age of 17 is substituted for the ages of 18 and 21 respectively in the case of—

    (a) motor vehicles owned or operated by the Secretary of State for Defence, or

    (b) motor vehicles driven by persons for the time being subject to the orders of a member of the armed forces of the Crown,

when they are being used for naval, military or air force purposes.

(11) In item 7, in the case of an incomplete large vehicle—

    (a) which has a working weight not exceeding 3.5 tonnes, the age of 17 is substituted for the age of 21;

    (b) which has a working weight exceeding 3.5 tonnes but not exceeding 7.5 tonnes, the age of 18 is substituted for the age of 21.

(**a**) 1992 c. 4.
(**b**) 1990 c. 19.
(**c**) 1978 c. 29.

(12) In item 7, the age of 17 is substituted for the age of 21 in the case of a road roller which—

    (a) is propelled otherwise than by steam,

    (b) has no wheel fitted with pneumatic, soft or elastic tyres,

    (c) has an unladen weight not exceeding 11.69 tonnes, and

    (d) is not constructed or adapted for the conveyance of a load other than the following things, namely water, fuel or accumulators used for the purpose of the supply of power to or propulsion of the vehicle, loose tools and objects specially designed to be attached to the vehicle for the purpose of temporarily increasing its weight.

(13) In this regulation—

    (a) for the purposes of paragraph (3)—

        (i) any implement fitted to a tractor shall be deemed to form part of the tractor notwithstanding that it is not a permanent or essentially permanent fixture,

        (ii) "closed-coupled", in relation to wheels on the same side of a trailer, means fitted so that at all times while the trailer is in motion the wheels remain parallel to the longitudinal axis of the trailer and that the distance between the centres of their respective areas of contact with the road surface does not exceed 840 millimetres, and

        (iii) "overall width", in relation to a vehicle, means the width of the vehicle measured between vertical planes parallel to the longitudinal axis of the vehicle and passing through the extreme projecting points thereof exclusive of any driving mirror and so much of the distortion of any tyre as is caused by the weight of the vehicle;

    (b) for the purposes of paragraph (8), "registered", "training agreement" and "the Training Scheme" have the meanings respectively given in regulation 54;

    (c) in paragraph (9), expressions used which are also used in Council Regulation 3820/85/EEC(a) have the same meanings as in that Regulation.

Applications for licences

Applications for the grant of licences: general

10.—(1) The Secretary of State may consider an application for the grant of a licence before the date on which the grant of the licence is to take effect if the application is received by him—

    (a) in the case of an application for a Group 2 licence, during the period of three months ending on that date,

    (b) in any other case, during the period of two months ending on that date,

and may during such period grant the licence so that it takes effect on that date.

(2) For the purposes of subsection (1A)(b) of section 89 of the Traffic Act the holder of an exchangeable licence satisfies the relevant residence requirement if he has been normally resident in Great Britain for a period of not more than five years.

(3) An applicant for a licence who before the licence is granted is required to satisfy the Secretary of State that he has passed a test shall at the time when he applies for the licence deliver to the Secretary of State—

    (a) a valid test pass certificate, or

    (b) a certificate corresponding to that certificate furnished under the law of Northern Ireland or Gibraltar.

(4) A person may not present a certificate in support of an application as evidence that he has passed—

    (a) a test or a theory test, or

    (b) a test corresponding to any of those tests conducted under the law of Northern Ireland or the law of Gibraltar,

---

(a) OJ No. L370, 31.12.85, p. 1. See also regulation 4 of the Community Drivers Hours and Recording Equipment (Exemptions and Supplementary Provisions) Regulations 1986 (S.I. 1986/1456).

if the applicant took the test in respect of which the certificate was issued at a time when he was ineligible, by virtue of an enactment contained in the Traffic Act or these Regulations or a corresponding provision of the law of Northern Ireland or the law of Gibraltar, to take the test to which the certificate relates.

(5) An applicant for a Group 2 licence shall, if required to do so by the Secretary of State, submit in support of his application a report (in such form as the Secretary of State may require) signed by a qualified medical practitioner, prepared and dated not more than four months prior to the date on which the licence is to take effect, for the purpose of satisfying the Secretary of State that he is not suffering from a relevant or prospective disability.

Eligibility to apply for provisional licence

11.—(1) Subject to the following provisions of this regulation, an applicant for a provisional licence authorising the driving of motor vehicles of a class included in a category or sub-category specified in column (1) of the table at the end of this regulation must hold a relevant full licence authorising the driving of motor vehicles of a class included in the category or sub-category specified in column (2) of the table in relation to the first category.

(2) Paragraph (1) shall not apply in the case of an applicant who is a full-time member of the armed forces of the Crown.

(3) For the purposes of paragraph (1), a licence authorising the driving only of vehicles in sub-categories D1 (not for hire or reward), D1 + E (not for hire or reward) and C1 + E (8.25 tonnes) shall not be treated as a licence authorising the driving of motor vehicles of a class included in sub-categories D1, D1 + E and C1 + E.

(4) In this regulation, "relevant full licence" means a full licence granted under Part III of the Traffic Act, a full Northern Ireland licence, a full British external licence (other than a licence which is to be disregarded for the purposes of section 89(1)(d) of the Traffic Act by virtue of section 89(2)(c) of that Act(**a**)), a full British Forces licence, an exchangeable licence or a Community licence.

TABLE

(1) Category or sub-category of licence applied for	(2) Category/sub-category of full licence required
B + E	B
C	B
C1	B
D	B
D1	B
C1 + E	C1
C + E	C
D1 + E	D1
D + E	D
G	B
H	B

Restrictions on the grant of large goods and passenger-carrying vehicle driver's licences

12.—(1) An applicant for a large goods or passenger-carrying vehicle driver's licence shall not, subject to paragraph (2), be granted a licence if, at the date from which the licence applied for is to take effect, any—

(a) large goods or passenger-carrying vehicle driver's licence held by him is suspended, or

(b) Northern Ireland large goods or passenger-carrying vehicle driver's licence held by him is suspended,

under section 115 of the Traffic Act or, as the case may be, under the provision of the law for the time being in force in Northern Ireland corresponding to that enactment.

(**a**) Subsection (2)(c) was substituted by the 1989 Act, section 4(3). For designations made under that provision see S.I. 1996/3206.

(2) A person may apply for a large goods vehicle driver's licence notwithstanding that, at the date from which the licence applied for is to take effect, any passenger-carrying vehicle driver's licence held by him is suspended and such suspension relates to his conduct other than as a driver of a motor vehicle.

(3) An applicant for an LGV trainee driver's licence—

(a) must be a registered employee of a registered employer (within the meaning of regulation 54), and

(b) must not be a person who—

(i) has been convicted (or is to be treated as if he had been convicted) of an offence as a result of which at least one penalty point falls to be taken into account under section 29 of the Offenders Act, or

(ii) has at any time been disqualified by a court for holding or obtaining a licence or by a court in Northern Ireland for holding or obtaining a Northern Ireland licence, and

(c) must satisfy the Secretary of State that he holds a Certificate of Professional Competence issued by the Road Haulage and Distribution Training Council stating that the applicant has completed a course of induction training in the driving of goods vehicles which meets the requirements of Council Directive 76/914/EEC(a).

(4) An applicant for a large goods vehicle driver's licence who is a member of the armed forces and is under the age of 21 must not be a person who has—

(a) been convicted (or is, by virtue of section 58 of the Offenders Act(b), to be treated as if he had been convicted) of an offence as a result of which at least one penalty point falls to be taken into account under section 29 of the Offenders Act(c), or

(b) at any time been disqualified by a court for holding or obtaining a licence or by a court in Northern Ireland for holding or obtaining a Northern Ireland licence.

Restrictions on the grant of provisional licences to drive motor bicycles

13.—(1) Subject to paragraphs (2) to (4), the Secretary of State must refuse to grant a provisional licence authorising the driving of a motor bicycle of any class to a person who was the holder of a previous licence if the licence applied for would come into force within the period of one year beginning on the day after the expiration of the period for which the previous licence authorised (or would, if not surrendered or revoked, have authorised) the riding of a motor bicycle.

(2) In a case where the applicant's previous licence was surrendered or revoked under subsection (3) or (4) of section 99 of the Traffic Act before its expiry date, paragraph (1) shall not apply.

(3) In a case where—

(a) the applicant's previous licence was surrendered or revoked, otherwise than under subsection (3) or (4) of section 99 of the Traffic Act, or treated as being revoked under section 37(1) of the Offenders Act, and

(b) the circumstances mentioned in regulation 15(2)(b) and (c) apply (so that the Secretary of State is required to grant a licence which would be in force for a period of less than two years),

the Secretary of State must refuse to grant a provisional licence which would come into force within the period of two months commencing on the date of such surrender or revocation.

(4) In a case where—

(a) the applicant's previous licence was surrendered or revoked, otherwise than under subsection (3) or (4) of section 99 of the Traffic Act, or treated as being revoked under section 37(1) of the Offenders Act, and

---

(a) OJ No. L357, 29.12.76, p. 36.
(b) Section 58(1) was amended by the 1990 Regulations.
(c) Section 29 was substituted by section 28 of the 1991 Act.

(b) the circumstances mentioned in regulation 15(2)(b) and (c) do not apply,

the Secretary of State must refuse to grant a provisional licence which would come into force within the period of one year commencing on the date of such surrender or revocation.

Fees for licences

14.—(1) An applicant for a licence shall pay a fee (if any) determined in accordance with paragraphs (2) and (3).

(2) The fee payable upon an application for a licence shall, in the case of a licence of a description, and (as the case may be) in the circumstances, specified in column (1) of the table set out in Schedule 3, be the fee specified in relation to that licence in column (2) of that table.

(3) When an application is made for a licence which, but for this paragraph, would attract more than one fee, only one fee shall be paid and where the fees are different, that fee shall be the higher or the highest of them.

Provisional licences

Duration of provisional licences authorising the driving of motor bicycles

15.—(1) Subject to paragraph (2), there is prescribed for the purposes of section 99(2) of the Traffic Act—

    (a) a motor bicycle of any class, and

    (b) a period of two years.

(2) There are prescribed for the purposes of section 99(2)(b)(ii) of that Act the circumstances that—

    (a) the previous licence was surrendered or revoked, otherwise than under subsection (3) or (4) of section 99 of the Traffic Act, or treated as being revoked under section 37(1) of the Offenders Act,

    (b) if it has not been so surrendered or revoked, a period of at least one month, commencing on the date of surrender or revocation, would have elapsed before the previous licence would have expired, and

    (c) the licence when granted would come into force within the period of one year beginning on the date of surrender or revocation of the previous licence.

Conditions attached to provisional licences

16.—(1) A provisional licence of any class is granted subject to the conditions prescribed in relation to a licence of that class in the following paragraphs.

(2) Subject to the following paragraphs, the holder of a provisional licence shall not drive a vehicle of a class which he is authorised to drive by virtue of that licence—

    (a) otherwise than under the supervision of a qualified driver who is present with him in or on the vehicle,

    (b) unless a distinguishing mark in the form set out in Part 1 of Schedule 4 is displayed on the vehicle in such manner as to be clearly visible to other persons using the road from within a reasonable distance from the front and from the back of the vehicle, or

    (c) while it is being used to draw a trailer.

(3) The condition specified in paragraph (2)(a) shall not apply when the holder of the provisional licence—

    (a) is driving a motor vehicle of a class included in sub-category B1 or B1 (invalid carriages) or in category F, G, H or K which is constructed to carry only one person and not adapted to carry more than one person;

    (b) is riding a moped or a motor bicycle with or without a side-car; or

    (c) is driving a motor vehicle, other than a vehicle of a class included in category C, C + E, D or D + E, on a road in an exempted island.

(4) The condition specified in paragraph (2)(b) shall not apply—

(a) when the holder of the provisional licence is driving a motor vehicle on a road in Wales, and

(b) a distinguishing mark in the form set out in Part 2 of Schedule 4 is displayed on the motor vehicle in the manner described in paragraph (2)(b).

(5) The condition specified in paragraph (2)(c) shall not apply to the holder of a provisional licence authorising the driving of a vehicle of a class included in category B + E, C + E, D + E or F, in relation to motor vehicles of that class.

(6) The holder of a provisional licence authorising the driving of—

(a) a moped, or

(b) a motor bicycle with or without a side-car,

shall not drive such a vehicle while carrying on it another person.

(7) The holder of a provisional licence authorising the driving of a motor bicycle other than a learner motor bicycle shall not drive such a vehicle otherwise than under the supervision of a certified direct access instructor (within the meaning of regulation 64(2)) who is—

(a) present with him on the road while riding another motor bicycle,

(b) able to communicate with him by means of a radio which is not hand-held while in operation,

(c) supervising only that person or only that person and another person who holds such a provisional licence, and

(d) carrying a valid certificate issued in respect of him by the Secretary of State under regulation 65(4),

while he and the instructor are wearing apparel which is fluorescent or (during hours of darkness) is either fluorescent or luminous.

(8) The holder of a passenger-carrying vehicle driver's provisional licence shall not drive a vehicle which he is authorised to drive by that licence while carrying any passenger in the vehicle other than—

(a) the person specified in paragraph (2)(a), or

(b) a person who holds a passenger-carrying vehicle driver's licence and either is giving or receiving instruction in the driving of passenger-carrying vehicles, or has given or received or is to give or receive, such instruction.

(9) The conditions specified in paragraphs (2)(a), (7) and (8) shall not apply when the holder of the provisional licence is undergoing a test.

(10) The conditions specified in paragraphs (2), (6), (7) and (8) shall not apply in relation to the driving of motor vehicles of a class in respect of which the provisional licence holder has been furnished with a valid test pass certificate stating that he has passed a test for the grant of a licence authorising him to drive vehicles of that class.

(11) The condition specified in paragraph (7)(b) shall not apply in the case of a provisional licence holder who is unable, by reason of impaired hearing, to receive directions from the supervising instructor by radio where the licence holder and the instructor are employing a satisfactory means of communication which they have agreed before the start of the journey.

(12) In the case of an LGV trainee driver's licence issued as a provisional licence, this regulation shall apply as modified by regulation 54.

(13) In this regulation—

(a) "exempted island" means any island outside the mainland of Great Britain from which motor vehicles, unless constructed or adapted specially for that purpose, cannot at any time be conveniently driven to a road in any other part of Great Britain by reason of the absence of any bridge, tunnel, ford or other way suitable for the passage of such motor vehicles but excluding any of the following islands, namely, the Isle of Wight, St. Mary's (Isles of Scilly), the islands of Arran, Barra, Bute, Great Cumbrae, Islay, the island which comprises Lewis and Harris, Mainland Orkney, Mainland Shetland, Mull, the island which comprises North Uist, Benbecula and South Uist and Tiree;

(b) "provisional licence", in relation to a class of vehicles, includes a full licence which is treated, by virtue of section 98 of the Traffic Act, as authorising its holder to drive vehicles of that class as if he held a provisional licence therefor;

(c) "qualified driver" shall be interpreted in accordance with regulation 17.

Meaning of "qualified driver"

17.—(1) Subject to paragraph (2), a person is a qualified driver for the purposes of regulation 16 if he—

(a) is 21 years of age or over,

(b) holds a relevant licence,

(c) has the relevant driving experience, and

(d) in the case of a disabled driver, he is supervising a provisional licence holder who is driving a vehicle of a class included in category B and would in an emergency be able to take control of the steering and braking functions of the vehicle in which he is a passenger.

(2) In the case of a person who is a member of the armed forces of the Crown acting in the course of his duties for naval, military or air force purposes sub-paragraphs (a) and (c) of paragraph (1) shall not apply.

(3) For the purposes of this regulation—

(a) "disabled driver" means a person who holds a relevant licence which is limited by virtue of a declaration made with his application for the licence or a notice served under section 92(5)(b) of the Traffic Act to vehicles of a particular class;

(b) "full licence" includes a full Northern Ireland licence and a Community licence;

(c) "relevant licence" means—

(i) in the case of a disabled driver, a full licence authorising the driving of a class of vehicles in category B other than vehicles in sub-category B1 or B1 (invalid carriages), and

(ii) in any other case, a full licence authorising the driving of vehicles of the same class as the vehicle being driven by the provisional licence holder;

(d) a person has relevant driving experience if he satisfies either of the following requirements—

(i) he has held the relevant licence for a continuous period of not less than 3 years or for periods amounting in aggregate to not less than 3 years, or

(ii) he is supervising a provisional licence holder driving a vehicle in category C, D, C + E or D + E, held the relevant licence on 6th April 1998, has held it continuously since that date and has held a full licence authorising the driving of vehicles in category B for a continuous period of not less than 3 years or for periods amounting in aggregate to not less than 3 years; and

(e) for the purposes of sub-paragraph (d)(ii) a person shall be regarded as holding a relevant licence during any period in which he holds a provisional licence and a valid test pass certificate entitling him to a full licence authorising the driving of vehicles of the same class as the vehicle being driven by the provisional licence holder.

Conditions attached to provisional licences: holders of driving permits other than licences granted under Part III of the Traffic Act

18. A holder of a provisional licence authorising the driving of vehicles of any class who also holds a permit by virtue of which he is at any time—

(a) treated, by virtue of regulation 80, as the holder, for the purposes of section 87 of the Traffic Act, of a licence authorising the driving of vehicles of that class, or

(b) entitled, pursuant to article 2(1) of the Motor Vehicles (International Circulation) Order 1975(a), to drive motor vehicles of that class,

need not comply with regulation 16 at that time.

Full licences not carrying provisional entitlement

19.—(1) The application of sections 98(2) and 99A(5) of the Traffic Act is limited or excluded in accordance with the following paragraphs.

(2) Subject to paragraphs (3), (4), (5), (6), (11) and (12), the holder of a full licence which authorises the driving of motor vehicles of a class included in a category or sub-category specified in column (1) of the table at the end of this regulation may drive motor vehicles—

    (a) of other classes included in that category or sub-category, and

    (b) of a class included in each category or sub-category specified, in relation to that category or sub-category, in column (2) of the table,

as if he were authorised by a provisional licence to do so.

(3) Section 98(2) shall not apply to a full licence if it authorises the driving only of motor vehicles adapted on account of a disability, whether pursuant to an application in that behalf made by the holder of the licence or pursuant to a notice served under section 92(5)(b) of the Traffic Act.

(4) In the case of a full licence which authorises the driving of a class of standard motor bicycles, other than bicycles included in sub-category A1, section 98(2) shall not apply so as to authorise the driving of a large motor bicycle by a person under the age of 21 before the expiration of the standard access period.

(5) In the case of a full licence which authorises the driving of motor bicycles of a class included in sub-category A1 section 98(2) shall not apply so as to authorise the driving of a large motor bicycle by a person under the age of 21.

(6) In the case of a full licence which authorises the driving of a class of vehicles included in category C or C + E, paragraph (2) applies subject to the provisions of regulation 54.

(7) Subject to paragraphs (8), (9), (10), (11) and (12), the holder of a Community licence to whom section 99A(5) of the Traffic Act applies and who is authorised to drive in Great Britain motor vehicles of a class included in a category or sub-category specified in column (1) of the Table at the end of this regulation may drive motor vehicles—

    (a) of other classes included in that category or sub-category, and

    (b) of a class included in each category or sub-category specified, in relation to that category or sub-category, in column (2) of the Table,

as if he were authorised by a provisional licence to do so.

(8) Section 99A(5) shall not apply to a Community licence if it authorises the driving only of motor vehicles adapted on account of a disability.

(9) In the case of a Community licence which authorises the driving of a class of standard motor bicycle other than bicycles included in sub-category A1, section 99A(5) shall not apply so as to authorise the driving of a large motor bicycle by a person under the age of 21 before the expiration of the period of two years commencing on the date when that person passed a test for a licence authorising the driving of that class of standard motor bicycle (and in calculating the expiration of that period, any period during which that person has been disqualified for holding or obtaining a licence shall be disregarded).

(10) In the case of a Community licence which authorises the driving only of motor bicycles of a class included in sub-category A1 section 98(2) shall not apply so as to authorise the driving of a large motor bicycle by a person under the age of 21.

(11) Except to the extent provided in paragraph (12), section 98(2) shall not apply to a full licence, and section 99A(5) shall not apply to a Community licence, in so far as it authorises its holder to drive motor vehicles of any class included in category B + E, C + E, D + E or K or in sub-category B1 (invalid carriages), C1 or D1 (not for hire or reward).

(a) S.I. 1975/1208. Article 2(1) was substituted by S.I. 1989/993 and amended by S.I. 1991/771.

(12) A person—

    (a) who holds a full licence authorising the driving only of those classes of vehicle included in a category or sub-category specified in paragraph (11) which have automatic transmission (and are not otherwise adapted on account of a disability), or

    (b) who holds a Community licence, to whom section 99A(5) of the Traffic Act applies and who is authorised to drive in Great Britain only those classes of vehicle included in a category or sub-category specified in paragraph (11) which have automatic transmission (and are not otherwise adapted on account of a disability),

may drive motor vehicles of all other classes included in that category or sub-category which have manual transmission as if he were authorised by a provisional licence to do so.

TABLE

(1) Full licence held	(2) Provisional entitlement included
A1	A, B, F and K
A	B and F
B1	A, B and F
B	A, B + E, G and H
C	C1 + E, C + E
D1	D1 + E
D	D1 + E, D + E
F	B and P
G	H
H	G
P	A, B, F and K

Miscellaneous

Signatures on licences

  20. In order that a licence may show the usual form of signature of its holder—

    (a) where the Secretary of State so requires, a person applying for a licence shall provide the Secretary of State with a specimen of his signature which can be electronically recorded and reproduced on the licence;

    (b) where no such requirement is made, a person to whom a licence is granted shall forthwith sign it in ink in the space provided.

Lost or defaced licences

  21.—(1) If the holder of a licence—

    (a) satisfies the Secretary of State that—

       (i) the licence or its counterpart has been lost or defaced; and

       (ii) the holder is entitled to continue to hold the licence; and

    (b) pays the fee prescribed by regulation 14,

the Secretary of State shall, on surrender of any licence or counterpart that has not been lost, issue to him a duplicate licence and counterpart and shall endorse upon the counterpart any particulars endorsed upon the original licence or counterpart as the case may be and the duplicates so issued shall have the same effect as the originals.

(2) If at any time while a duplicate licence is in force the original licence is found, the person to whom the original licence was issued, if it is in his possession, shall return it to the Secretary of State, or if it is not in his possession, but he becomes aware that it is found, shall take all reasonable steps to take possession of it and if successful shall return it as soon as may be to the Secretary of State.

(3) The obligation in paragraph (2) shall apply in respect of the counterpart of a licence as if for the words "original licence" in each place where they occur there were substituted the words "original counterpart".

PART III

TESTS OF COMPETENCE TO DRIVE

Preliminary

Interpretation of Part III

22. In this Part of these Regulations—

"applicant in person" means a person making an application for an appointment for a test or a part of a test with a view to taking the test or that part thereof himself;

"appointed person" means a person appointed by the Secretary of State to conduct theory tests under paragraph (1)(a)(ii) or (2)(a) of regulation 23;

"DSA examiner" means a person appointed by the Secretary of State to conduct practical or unitary tests under paragraph (1)(a) or (2)(a) of regulation 24;

"large vehicle instructor" means a person operating an establishment for providing instruction in the driving of vehicles included in category B + E, C, C + E, D or D + E, including an establishment which provides tuition to prepare persons for the theory test;

"motor bicycle instructor" means a person operating an establishment for providing instruction in the driving of vehicles included in categories A or P, including an establishment which provides tuition to prepare persons for the theory test;

"standard access period" means the period of two years commencing on the date when a person passes a test for a licence authorising the driving of standard motor bicycles of any class, other than a class included in the sub-category A1, but disregarding—

(a) any period during which the person is disqualified under section 34(**a**) or 35(**b**) of the Offenders Act,

(b) in a case where the person has been disqualified under section 36 of the Offenders Act(**c**), the period beginning on the date of the court order under subsection (1) of that section and ending on the date when the disqualification is deemed by virtue of that section to have expired in relation to standard motor bicycles of that class,

(c) in a case where the Secretary of State has revoked the person's licence or test pass certificate under section 3(2) of, or Schedule 1 to, the Road Traffic (New Drivers) Act 1995(**d**), the period beginning on the date of the notice of revocation under that Act and ending on the date when the person passes the relevant driving test within the meaning of that Act, and

(d) any period during which the licence has ceased to be in force;

"working day" means a day other than a Saturday, Sunday, bank holiday, Christmas Day or Good Friday (and "bank holiday" means a day to be observed as such under section 1 of and Schedule 1 to the Bank and Financial Dealings Act 1971(**e**)).

Appointment of persons to conduct tests

Persons by whom theory tests may be conducted

23.—(1) A theory test other than a test conducted in the circumstances specified in paragraph (2) may be conducted by—

(a) a person appointed in writing by the Secretary of State—

(i) for the purpose of testing a class of persons specified in the instrument of appointment, or

(ii) where no class of persons is specified, for the purpose of testing persons generally;

(**a**) Amended by section 29 of the 1991 Act and section 3(2) of the Aggravated Vehicle-Taking Act 1992 (c. 11).

(**b**) Amended by paragraph 95 of Schedule 2 of the 1991 Act.

(**c**) Section 36 was substituted by section 32 of the 1991 Act and amended by the 1996 Regulations and the Deregulation (Exchangeable Driving Licences) Order 1998 (S.I. 1998/1917).

(**d**) 1995 c. 13.

(**e**) 1971 c. 80.

(b) a person who, or a member of a class of persons which, has been appointed by the Secretary of State for Defence, for the purpose of testing members of the armed forces of the Crown and persons in the public service of the Crown under his department;

(c) a person appointed by a chief officer of police, for the purpose of testing—

    (i) members of the police force of which he is the chief officer and persons employed by the police authority for the same police area for the purpose of assisting that force, and,

    (ii) members of another police force and persons employed by a police authority for another police area for the purpose of assisting that force;

(d) in England and Wales, a person appointed by the chief officer of any fire brigade maintained in pursuance of the Fire Services Act 1947(**a**) or, in Scotland, by the firemaster of such a brigade, for the purpose of testing members of the brigade of which he is the chief officer or of persons employed in the driving of motor vehicles for the purposes of any such brigade;

(e) an eligible person appointed by a company which—

    (i) has been approved by the Secretary of State, and

    (ii) is the holder of a PSV operator's licence,

for the purpose of conducting, in respect of eligible candidates, theory tests in respect of any class of passenger-carrying vehicles.

(2) Where the person submitting himself for a test is disqualified until he passes the appropriate driving test, a theory test shall be conducted by—

(a) a person appointed by the Secretary of State for the purpose;

(b) a person who, or a member of a class of persons which, has been appointed by the Secretary of State for Defence, for the purpose of testing members of the armed forces of the Crown and persons in the public service of the Crown under his department.

(3) No person shall be eligible to appoint any person or class of persons to conduct theory tests under the provisions of sub-paragraphs (b), (c), (d) or (e) of paragraph (1) or under paragraph (2)(b) unless, following an application made to him for the purpose of any of those sub-paragraphs, the Secretary of State is satisfied that—

(a) proper arrangements will be made by the applicant, for the conduct of tests in accordance with these Regulations; and

(b) proper records of such tests and the results thereof will be kept by him or them,

and has granted his approval in writing, subject to such conditions as he thinks fit to impose.

(4) In the case of an application made by a chief officer of police for the purposes of sub-paragraph (c) of paragraph (1), the Secretary of State may grant his approval under paragraph (3) in respect of the testing of all the persons mentioned in that sub-paragraph or only in respect of the testing of the persons mentioned in paragraph (i) thereof.

(5) No person or class of persons may be appointed under the provisions of paragraph (b), (c), (d) or (e) of paragraph (1) or under paragraph (2)(b) unless the person making the appointment has appointed a person or class of persons to conduct practical tests under the provisions of regulation 24(1) and the Secretary of State has approved that appointment.

(6) An appointment made under paragraph (1)(a)(ii) may be made subject to such conditions as are, in the opinion of the Secretary of State, reasonably necessary in the general interests of candidates and where an appointed person breaks such a condition the Secretary of State may appoint another person to carry out theory tests in substitution for that person notwithstanding that the first appointment has not been revoked.

(7) A person may not conduct a test prescribed in respect of any category or sub-category of motor vehicle unless he is expressly appointed for the purpose of conducting such a test.

(**a**) 1947 c. 41.

(8) No person or member of a class of persons appointed by virtue of sub-paragraph (b), (c), (d) or (e) of paragraph (1) or under paragraph (2)(b) may conduct tests unless the Secretary of State has given his approval in writing to the appointment and such approval shall be granted only if the Secretary of State is satisfied that the person (or, in the case of the appointment of a class of persons, each member of that class) is capable of making a proper assessment of a candidate's knowledge and understanding of driving theory relating to the category or sub-category of vehicles in respect of which he is appointed to conduct tests.

(9) In this regulation and regulation 24—

"chief officer of police", "police area" and "police authority" have the meanings given in section 101(1) of the Police Act 1996(**a**);

"company" includes a body corporate;

"eligible candidate" means—

(i) a person who is employed as a driver by the company which holds the PSV operator's licence or by a sister company of that company which also holds a PSV operator's licence, or

(ii) a person whom any such company as is mentioned in sub-paragraph (i) proposes to employ as a driver;

"eligible person" means a person employed by the company which hold the PSV operator's licence or by a sister company of that company if the sister company also holds a PSV operator's licence,

and a company is a sister company of another if either is the holding company of the other or both are wholly-owned subsidiaries of a third within the meaning of section 736 of the Companies Act 1985(**b**).

Persons by whom practical and unitary tests may be conducted

24.—(1) A practical or unitary test other than a test conducted in the circumstances specified in paragraph (2) may, subject to the following provisions of this regulation, be conducted by—

(a) a person in the public service of the Crown appointed by the Secretary of State;

(b) a person who, or a member of a class of persons which, has been appointed by the Secretary of State for Defence, for the purpose of testing members of the armed forces of the Crown and persons in the public service of the Crown under his department;

(c) in England and Wales, a person appointed by the chief officer of any fire brigade maintained in pursuance of the Fire Services Act 1947 or, in Scotland, by the firemaster of such a brigade, for the purpose of testing members of the brigade of which he is the chief officer or of persons employed in the driving of motor vehicles for the purposes of any such brigade;

(d) a person appointed by a chief officer of police, for the purpose of testing—

(i) members of the police force of which he is the chief officer and persons employed by the police authority for the same police area for the purpose of assisting that force, and

(ii) members of another police force and persons employed by a police authority for another police area for the purpose of assisting that force;

(e) a person appointed by a company which—

(i) has been approved by the Secretary of State, and

(ii) normally employs for the purpose of its operations in excess of 250 persons as drivers of motor vehicles,

for the purpose of testing persons employed by it as drivers or persons whom it proposes so to employ;

(f) an eligible person appointed by a company which—

(i) has been approved by the Secretary of State, and

(ii) is the holder of a PSV operator's licence,

for the purpose of conducting, in respect of eligible candidates, practical tests in respect of any class of passenger-carrying vehicles.

(**a**) 1996 c. 16.
(**b**) 1985 c. 6. Section 736 was substituted by the Companies Act 1989, section 144(1).

(2) Where the person submitting himself for a test is disqualified until he passes the appropriate driving test, a practical or unitary test shall be conducted by—

(a) a person in the public service of the Crown appointed by the Secretary of State, or

(b) a person who, or a member of a class of persons which, has been appointed by the Secretary of State for Defence, for the purpose of testing members of the armed forces of the Crown and persons in the public service of the Crown under his department.

(3) No person shall be eligible to appoint any person or class of persons to conduct practical or unitary tests under the provisions of sub-paragraphs (b), (c), (d), (e) or (f) of paragraph (1) or under paragraph (2)(b) unless, following an application made to him for the purpose of any of those sub-paragraphs, the Secretary of State is satisfied that—

(a) proper arrangements will be made by the applicant, for the conduct of tests in accordance with these Regulations; and

(b) proper records of such tests and the results thereof will be kept by him or them,

and has granted his approval in writing, subject to such conditions as he thinks fit to impose.

(4) In the case of an application made by a chief officer of police for the purposes of sub-paragraph (d) of paragraph (1), the Secretary of State may grant his approval under paragraph (3) in respect of the testing of all the persons mentioned in that sub-paragraph or only in respect of the testing of the persons mentioned in paragraph (i) thereof.

(5) No person or member of a class of persons appointed under the provisions of sub-paragraph (b), (c), (d), (e) or (f) of paragraph (1) or under paragraph (2)(b) may conduct tests unless the Secretary of State has given his approval in writing to his appointment and such approval shall be granted only if the Secretary of State is satisfied that the person (or, in the case of the appointment of a class of persons, each member of that class) is capable of making a proper assessment of a candidate's ability to drive vehicles of the class in respect of which he is appointed to conduct tests.

(6) A person may not conduct a test prescribed in respect of any category or sub-category of motor vehicle unless he is expressly appointed for the purpose of conducting such a test.

Revocation of authority to conduct tests

25.—(1) The Secretary of State may revoke—

(a) an appointment made under regulation 23(1)(a) or (2)(a) or under regulation 24(1)(a) or (2)(a), or

(b) an approval given under regulation 23(3) or (8) or under regulation 24(3) or (5),

by notice in writing and the authority of the person whose appointment is revoked or whose approval is withdrawn to conduct theory tests or, as the case may be, to appoint other persons to conduct unitary, practical or theory tests, shall cease upon the date specified in the notice.

(2) Where a person has his appointment revoked or if an approval given in respect of him under regulation 23(3) or 24(3) is withdrawn, that person shall immediately return to the Secretary of State all forms of pass certificates supplied to him under regulations 47(8) and 48(3) which he still holds.

Applications for tests

Applications for theory tests: applicants in person

26.—(1) An applicant in person wishing to take a theory test to be conducted by an appointed person shall—

(a) apply for an appointment to that person,

(b) provide that person with such details relating to himself, the licence which he holds, the preferred location of the test, and the nature of the test to be taken as he may reasonably require, and

(c) in the case of an application for a test to be conducted before 4th January 2000, state whether or not he requires the theory test pass certificate or failure statement to be furnished under regulation 47(2) on the day of the test and pay the fee specified in regulation 30.

(2) Upon receipt of such details and such fee the appointed person shall make the arrangements necessary for taking the theory test.

(3) An applicant in person for whom an appointment is made as aforesaid in respect of any category of motor vehicle may neither apply as an applicant in person nor be nominated by virtue of regulation 27 or 28 for a further appointment for a theory test in respect of the same category unless—

(a) the first appointment has been cancelled, or

(b) the test due on the first appointment does not take place for any reason other than cancellation; or

(c) he has kept the first appointment (whether or not the test is completed).

Applications for theory tests: motor bicycle instructors

27.—(1) A motor bicycle instructor who wishes to make an appointment for a theory test prescribed in respect of motor vehicles in category A or P to be conducted by an appointed person and to be taken by a person who has, or will have, received from that instructor tuition to prepare him for the theory test shall—

(a) apply for such an appointment to the appointed person, specifying the date and time of the appointment which the instructor wishes to reserve and the place where he wishes the test to be conducted,

(b) provide such details relating to himself, the establishment and the nature of the test as the appointed person may reasonably require,

(c) where the application proposes an appointment for a test on a date before 4th January 2000, state whether or not he requires the theory test pass certificate or failure statement to be furnished under regulation 47(2) to the person nominated under paragraph (4) on the day of the test, and

(d) pay the fee (recoverable from the person nominated under paragraph (4)) specified in regulation 30.

(2) The appointed person may refuse to accept an application from a motor bicycle instructor (or, where two or more applications have been made on the same occasion, to accept all or any of those applications) where any appointment specified in the application is unavailable or where, in the opinion of the appointed person, it is reasonably necessary to do so in the general interests of applicants for theory tests.

(3) Subject to paragraphs (2) and (5), upon receipt of such details and such fee the appointed person shall confirm to the motor bicycle instructor the date and time of the appointment.

(4) If, before the expiration of the qualifying period, the appointed person receives from the motor bicycle instructor the name and such further details relating to—

(a) the person receiving tuition from that instructor who will at the appointment submit himself for that test, and

(b) the nature of the test,

as the appointed person may reasonably require, the appointed person shall make the arrangements necessary for the taking of the appropriate test.

(5) A person nominated by a motor bicycle instructor pursuant to paragraph (4) for a theory test in respect of motor vehicles in category A or P may neither be so nominated nor apply under regulation 26 for a further appointment for such a test unless—

(a) the appointment made pursuant to the first nomination has been cancelled, or

(b) the test due on that appointment does not take place for any reason other than cancellation, or

(c) he has kept that appointment (whether or not the test is completed).

(6) The qualifying period for the purposes of paragraph (4) is the period expiring on the day which is three clear working days before the day for which the appointment is made.

Applications for theory tests: large vehicle instructors

28.—(1) A large vehicle instructor who wishes to make an appointment for a theory test prescribed in respect of motor vehicles in category C or D to be conducted by an appointed person and to be taken by a person who has, or will have, received from that instructor tuition to prepare him for the theory test shall—

(a) apply for such an appointment to the appointed person, specifying the date and time of the appointment which the instructor wishes to reserve and the place where he wishes the test to be conducted,

(b) provide such details relating to himself, the establishment and the nature of the test as the appointed person may reasonably require,

(c) where the application proposes an appointment for a test on a date before 4th January 2000, state whether or not he requires the theory test pass certificate or failure statement to be furnished under regulation 47(2) to the person nominated under paragraph (4) on the day of the test, and

(d) pay the fee (recoverable from the person nominated under paragraph (4)) specified in regulation 30.

(2) The appointed person may refuse to accept an application from a large vehicle instructor (or, where two or more applications have been made on the same occasion, to accept all or any of those applications) where any appointment specified in the application is unavailable, or where, in the opinion of the appointed person, it is reasonably necessary to do so in the general interests of applicants for theory tests.

(3) Subject to paragraphs (2) and (5), upon receipt of such details and such fee the appointed person shall confirm to the large vehicle instructor the date and time of the appointment.

(4) If, before the expiration of the qualifying period, the appointed person receives from the large vehicle instructor the name and such further details relating to—

(a) the person receiving tuition from that instructor who will at the appointment submit himself for that test, and

(b) the nature of the test,

as the appointed person may reasonably require, the appointed person shall make the arrangements necessary for the taking of the appropriate test.

(5) A person nominated by a large vehicle instructor pursuant to paragraph (4) for a theory test prescribed in respect of any category may neither be so nominated nor apply under regulation 26 for a further appointment for such a test unless—

(a) the appointment made pursuant to the first nomination has been cancelled, or

(b) the test due on that appointment does not take place for any reason other than cancellation, or

(c) he has kept that appointment (whether or not the test is completed).

(6) The qualifying period for the purposes of paragraph (4) is the period ending on the day which is three clear working days before the day for which the appointment is made.

Eligibility to reapply for theory test

29.—(1) Subject to paragraph (2), a person who has failed to pass a theory test prescribed in respect of any category may not make an application for another test of that nature to be conducted before the expiry of a period of three clear working days commencing with the day after the date of the first test.

(2) Paragraph (1) shall not apply—

(a) in a case where the person is either a member of the armed forces of the Crown or a person in the civil service of the Crown under the Secretary of State for Defence and the application is made with the consent of a person authorised by the Secretary of State for Defence; or

(b) in a case where the first test is conducted by an appointed person in accordance with paragraph (1)(a) or (2)(a) of regulation 23 and the Secretary of State has, prior to that test, given notice to the person that he will accept an application for a further test to be conducted before the expiry of the period mentioned in paragraph (1).

Fees for theory tests

30.—(1) The fee payable for a theory test to be conducted by an appointed person in respect of any category of motor vehicle is—

(a) in a case of an application for a theory test to be conducted before 4th January 2000 under which the theory test pass certificate or failure statement is required to be furnished under regulation 47(2) on the day of the test, £21, and

(b) in the case of any other application, £15.50,

and is payable to the Secretary of State.

(2) No fee is payable for a theory test conducted by any other person.

Applications for practical and unitary tests: applicants in person

31.—(1) An applicant in person wishing to take a practical or unitary test to be conducted by a DSA examiner shall—

(a) apply for an appointment for such a test to the Secretary of State,

(b) provide the Secretary of State with such details relating to himself, the licence which he holds, the preferred location of the test, the nature of the test and the vehicle on which the test is to be taken as the Secretary of State may reasonably require, and

(c) pay such fee as is specified in regulation 35.

(2) Upon receipt of such details and such fee the Secretary of State shall make the arrangements necessary for the taking of the appropriate test.

(3) An applicant in person for whom an appointment is made as aforesaid in respect of a class of motor vehicle in any category may neither apply as an applicant in person nor be nominated by virtue of regulation 32 or 33 for a further appointment for a practical or, as the case may be, a unitary test on a vehicle of the same class unless—

(a) the first appointment has been cancelled, or

(b) the test due on the first appointment does not take place for any reason other than cancellation, or

(c) he has kept the first appointment (whether or not the test is completed).

Applications for practical tests: motor bicycle instructors

32.—(1) A motor bicycle instructor who wishes to make an appointment for a practical test prescribed in respect of category A or P which is to be conducted by a DSA examiner and taken by a person who has, or will have, received from that instructor instruction in the driving of vehicles of a class included in either category shall—

(a) apply for such an appointment to the Secretary of State, specifying the date and time for the appointment which the instructor wishes to reserve and the place where he wishes the test to be conducted,

(b) provide such details relating to—

(i) himself,

(ii) his establishment,

(iii) the vehicle on which the test is to be taken, and

(iv) the nature of the test,

as the Secretary of State may reasonably require, and

(c) pay such fee (recoverable from the person nominated under paragraph (4)) as is specified in regulation 35.

(2) The Secretary of State may refuse to accept an application from a motor bicycle instructor (or, where two or more applications have been made on the same occasion, to accept all or any of those applications) where an appointment specified in the application is unavailable or where, in the opinion of the Secretary of State, it is reasonably necessary to do so in the general interests of applicants for practical or unitary tests.

(3) Subject to paragraphs (2) and (5), upon receipt of such details and such fee the Secretary of State shall confirm to the motor bicycle instructor the date and time of the appointment.

(4) If, before the expiration of the qualifying period, the Secretary of State receives from the motor bicycle instructor the name and such further details relating to—

(a) the person receiving instruction from that instructor who will at the appointment submit himself for that test,

(b) the licence which that person holds,

(c) the nature of the test, and

(d) the vehicle on which the test is to be taken,

as the Secretary of State may reasonably require, the Secretary of State shall make the arrangements necessary for the taking of the appropriate test.

(5) A person nominated by a motor bicycle instructor pursuant to paragraph (4) for a practical test in respect of any class of motor vehicle included in category A or P may neither be so nominated nor apply under regulation 31 for a further appointment for a test in respect of a motor vehicle of the same class unless—

(a) the appointment made pursuant to the first nomination has been cancelled, or

(b) the test due on that appointment does not take place for any reason other than cancellation, or

(c) he has kept that appointment (whether or not the test is completed).

(6) The qualifying period for the purposes of paragraph (4) is the period expiring at midday on the day which is two working days before the day for which the appointment is made.

Applications for practical tests: large vehicle instructors

33.—(1) A large vehicle instructor who wishes to make an appointment for a practical test prescribed in respect of category B + E, C, C + E, D or D + E which is to be conducted by a DSA examiner and taken by a person who has, or will have, received from that instructor instruction in the driving of a class of vehicle included in any of those categories shall—

(a) apply for such an appointment to the Secretary of State, specifying the date and time for the appointment which the instructor wishes to reserve and the place where he wishes the test to be conducted,

(b) provide such details relating to—

(i) himself,

(ii) his establishment,

(iii) the vehicle on which the test is to be taken, and

(iv) the nature of the test,

as the Secretary of State may reasonably require, and

(c) pay such fee (recoverable from the person nominated under paragraph (4)) as is specified in regulation 35.

(2) The Secretary of State may refuse to accept an application from a large vehicle instructor (or, where two or more applications have been made on the same occasion, to accept all or any of those applications) where an appointment specified in the application is unavailable or where, in the opinion of the Secretary of State, it is reasonably necessary to do so in the general interests of applicants for practical or unitary tests.

(3) Subject to paragraphs (2) and (5), upon receipt of such details and such fee the Secretary of State shall confirm to the large vehicle instructor the date and time of the appointment.

(4) If, before the expiration of the qualifying period, the Secretary of State receives from the large vehicle instructor the name and such further details relating to—

(a) the person receiving instruction from that instructor who will at the appointment submit himself for that test,

(b) the licence which that person holds,

(c) the nature of the test, and

(d) the vehicle on which the test is to be taken,

as the Secretary of State may reasonably require, the Secretary of State shall make the arrangements necessary for the taking of the appropriate test.

(5) A person nominated by a large vehicle instructor pursuant to paragraph (4) for a practical test in respect of any class of vehicle may neither be so nominated nor apply under regulation 31 for a further appointment for a test in respect of a motor vehicle of a class included in the same category unless—

(a) the appointment made pursuant to the first nomination has been cancelled, or

(b) the test due on that appointment does not take place for any reason other than cancellation, or

(c) he has kept that appointment (whether or not the test is completed).

(6) The qualifying period for the purposes of paragraph (4) is the period expiring at midday on the day which is two working days before the day for which the appointment is made.

Eligibility to reapply for practical or unitary test

34.—(1) Subject to the following provisions of this regulation, a person who has failed to pass a practical or unitary test ("the first test") for a licence authorising the driving of vehicles of a class included in any category may not make an application for another test for a licence authorising the driving of vehicles of any class included in the same category to be conducted before the expiry of the relevant period.

(2) Paragraph (1) shall not apply—

(a) in a case where the person is either a member of the armed forces of the Crown or a person in the civil service of the Crown under the Secretary of State for Defence and the application is made with the consent of a person authorised by the Secretary of State for Defence; or

(b) in a case where the first test is conducted by a DSA examiner and the Secretary of State has, prior to that test, given notice to the person that he will accept an application for a further test to be conducted before the expiry of the relevant period.

(3) In this regulation, "the relevant period" means—

(a) in the case of a test for a licence authorising the driving of a vehicle of a class included in category C, C + E, D or D + E, 3 clear working days, and

(b) in any other case, 10 clear working days.

commencing with the first day after the date of the first test.

Fees in respect of practical or unitary tests

35.—(1) No fee shall be payable in respect of a practical or unitary test conducted by a person appointed under regulation 24(1)(b), (c), (d), (e) or (f) or (2)(b).

(2) Subject to paragraphs (4) and (5), in the case of a practical or unitary test which—

(a) is to be conducted by a DSA examiner,

(b) is not, or does not form part of, an extended driving test,

(c) is for a licence authorising the driving of a motor vehicle of a class included in a category or sub-category specified in column (1) of the Table in Schedule 5,

the fee payable is that specified in relation to that category or sub-category in column (2) of that Table.

(3) Subject to paragaph (4), in the case of a practical or unitary test which—

(a) is to be conducted by a DSA examiner,

(b) is, or forms part of, an extended driving test,

(c) is for a licence authorising the driving of a motor vehicle of a class included in a category or sub-category specified in column (1) of the Table in Schedule 5,

the fee payable is that specified in relation to that category or sub-category in column (3) of that Table.

(4) Where an appointment for a practical test to commence during normal hours is cancelled by or on behalf of the Secretary of State and the appointment cannot reasonably be rearranged so that the test commences during normal hours, the applicant shall pay the fee prescribed for a test commencing during normal hours notwithstanding that it commences out of hours.

(5) In a case where the test is for a licence authorising the driving of vehicles included in category B and the applicant holds a full licence authorising the driving of vehicles included in sub-category B1 (invalid carriages), no fee shall be payable.

(6) For the purposes of this regulation and Schedule 5, a test commences—

(a) during normal hours, if the time for which the test appointment is made is any time between 0830 hours and 1630 hours on a working day, and

(b) out of hours, if the time for which the test appointment is made is any other time.

Cancellation of tests

36. For the purposes of paragraph (b) of section 91 of the Traffic Act (which section specifies the cases in which a fee paid on an application for an appointment for a test may be repaid) notice cancelling an appointment—

(a) for a practical or unitary test to be conducted by a DSA examiner must be given to the Secretary of State not less than ten clear working days before the day for which the appointment is made;

(b) for a theory test to be conducted by an appointed person must be given not less than three clear working days before the day for which the appointment is made.

Requirements at tests

Test vehicles

37.—(1) Subject to paragraph (3), the prescribed practical or unitary test for a licence authorising the driving of vehicles included in a category shown in column (1) of the Table at the end of this regulation must be conducted in a vehicle having a specification equivalent to or (except in the case of a test prescribed in respect of category F, G, H, K or P) higher than that shown in relation to that category in column (2) of the Table.

(2) Subject to paragraph (3), the prescribed practical test for a licence authorising the driving of vehicles included in a sub-category shown in column (1) of the Table at the end of this regulation must be conducted in a vehicle having a specification equivalent to or higher than that shown in relation to that sub-category in column (2) of the Table.

(3) Where an applicant for a practical test prescribed in respect of category A declares that he is suffering from a relevant disability of such a nature that he is unable to ride a motor bicycle without a side-car, that test must be conducted on a motor bicycle and side-car combination having the following specification—

(a) in the case of a test for a licence authorising the driving of a large motor bicycle and side-car combination, a combination in which the bicycle has a maximum net power output of not less than 35 kilowatts,

(b) in the case of a test for a licence authorising the driving of a standard motor bicycle and side-car combination (other than a combination included in sub-category A1), a combination which has a power to weight ratio not exceeding 0.16 kw/kg., and

(c) in the case of a test for a licence authorising the driving of a motor bicycle and side-car combination included in sub-category A1, a combination consisting of a minimum test vehicle for that sub-category and a side-car where the combination has a power to weight ratio not exceeding 0.16 kw/kg.

(4) A person submitting himself for a practical or unitary test shall provide a vehicle which—

(a) corresponds to the specification referred to in paragraph (1), (2) or (3), as the case may be,

(b) is not fitted with a device designed to permit a person other than the driver to operate the accelerator, unless any pedal or lever by which the device is operated and any other parts which it may be necessary to remove to make the device inoperable by such a person during the test have been removed, and

(c) is reasonably representative of the class to which it belongs and is otherwise suitable for the purposes of the test.

(5) A person submitting himself for a practical test prescribed in respect of category B or B + E shall provide a motor vehicle which—

(a) is fitted with a front passenger seat unless it—

(i) is a vehicle included in sub-category B1 and is constructed without a front passenger seat, or

(ii) has been adapted on account of a disability of the person who has submitted himself for the test and as part of that adaptation has had the front passenger seat removed,

(b) has fitted for use with the front passenger seat (or, if there is no such seat, with another seat in which the person conducting the test may conveniently sit for the purpose of the test) a properly anchored and functioning three-point seat belt, and

31

(c) in the case of a vehicle fitted with a front passenger seat, has fitted as an integral part of that seat a head restraint which satisfies the requirements of Council Directive 78/932/EEC(**a**).

(6) A person submitting himself for a practical test prescribed in respect of category B shall provide a vehicle which is fitted with an interior rear-view mirror providing adequate rearward vision from the front passenger seat unless it—

    (a) is a vehicle included in sub-category B1 and is constructed without a front passenger seat, or

    (b) has been adapted on account of a disability of the person who has submitted himself for the test and as part of that adaptation has had the front passenger seat removed.

(7) A person submitting himself for a practical test prescribed in respect of category B + E, C, C + E, D or D + E shall provide a motor vehicle which is not carrying goods or burden other than fixed items which are characteristic of the class to which it belongs.

(8) A person submitting himself for a practical test prescribed in respect of category C, C + E, D or D + E shall provide a motor vehicle which is fitted with a seat which is firmly secured to the vehicle and in such a position that the person conducting the test may properly do so and is protected from bad weather during the test.

(9) A person submitting himself for a practical test prescribed in respect of category D or D + E shall provide a motor vehicle which is fitted with a seat which is so placed that the person conducting the test can, from the deck of the vehicle on which the driver is seated, clearly observe the road to the rear of the vehicle without the use of any optical device, unless—

    (a) the construction of the vehicle makes it impossible to fulfil that requirement, or

    (b) the examiner consents to the requirement not being complied with in consequence of an arrangement to conduct part of the test elsewhere than on a road.

(10) A person submitting himself for a practical test prescribed in respect of category B + E, C + E or D + E shall provide a motor vehicle which is fitted with linkage and braking mechanisms which are designed for use when the trailer is fully laden.

(11) In the case of a test being conducted by a person appointed in accordance with paragraph (1)(b) or (2)(b) of regulation 24, paragraphs (5)(c) and (6) shall not apply.

(12) In the table at the end of this regulation, "minimum test vehicle" means, in relation to any category or sub-category, a vehicle of a specification shown in relation to the category or sub-category in column (2) of the table.

TABLE

(1) Category or sub-category	(2) Specification
A in the case of a test for a licence authorising the driving of large motor bicycles	A motor bicycle without a sidecar having an engine with a maximum net power output of 35 kilowatts.
A in the case of any other test	A learner motor bicycle without a sidecar having an engine with a cylinder capacity of 121 cubic centimetres and capable of a speed of 100 kilometres per hour.
A1	A learner motor bicycle without a sidecar having an engine with a cylinder capacity of 75 cubic centimetres.
B	Any four-wheeled vehicle in category B capable of a speed of 100 kilometres per hour.
B1	Any vehicle in sub-category B1 capable of a speed of 60 kilometres per hour.
B + E	A combination of a minimum test vehicle for category B and a trailer having a maximum authorised mass of 1,000 kilograms which is capable of a speed of 100 kilometres per hour.

(**a**) Council Directive of 16 October 1978 on the approximation of laws relating to the head restraints of seats of motor vehicles, OJ No. L325, 20.11.78, p. 1.

(1) Category or sub-category	(2) Specification
C1	Any vehicle in sub-category C1 having a maximum authorised mass of 4,000 kilograms and capable of a speed of 80 kilometres per hour.
C1 + E	A combination of a minimum test vehicle for sub-category C1 and a trailer having a maximum authorised mass of 2,000 kilograms, the overall length of which is 8 metres and which is capable of a speed of 80 kilometres per hour.
D1	Any vehicle in sub-category D1 capable of a speed of 80 kilometres per hour.
D1 + E	A combination of a minimum test vehicle for sub-category D1 and a trailer having a maximum authorised mass of 1,250 kilograms which is capable of a speed of 80 kilometres per hour.
C	Any vehicle in category C, other than an articulated goods vehicle, having a maximum authorised mass of 10,000 kilograms and a length of 7 metres which is capable of a speed of 80 kilometres per hour.
C + E	Either— (a) an articulated goods vehicle combination having a maximum authorised mass of 18,000 kilograms and a length of 12 metres which is capable of a speed of 80 kilometres per hour, or (b) a combination of a minimum test vehicle for category C and a trailer having a length of 4 metres and a maximum authorised mass of 4 tonnes, which has, in aggregate, a maximum authorised mass of 18,000 kilograms and an overall length of 12 metres and which is capable of a speed of 80 kilometres per hour.
D	Any vehicle in category D having a length of 9 metres and capable of a speed of 80 kilometres per hour.
D + E	A combination of a minimum test vehicle for category D and a trailer having a maximum authorised mass of 1,250 kilograms which is capable of a speed of 80 kilometres per hour.
F	Any vehicle in category F.
G	Any vehicle in category G.
H	Any vehicle in category H.
K	Any vehicle in category K.
P	Any vehicle in category P.

Further requirements at tests

38.—(1) Subject to paragraph (2), no person shall submit himself for a theory test, practical test or unitary test unless he satisfies the residence requirement specified in section 89(1A) of the Traffic Act and where any question arises as to whether a person is normally resident in Great Britain or the United Kingdom (as the case may be) he shall be deemed to be normally resident there if he shows that he will have lived there for not less than 185 days during the period of 12 months ending on the day for which the test appointment is made.

(2) Paragraph (1) shall not apply in the case of a person who submits himself for an appropriate driving test pursuant to section 36 of the Offenders Act or for any part of such a test.

(3) A person submitting himself for a theory test shall—

(a) before the test commences—

(i) except in a case to which paragraph (7) applies, produce to the person conducting the test an appropriate licence authorising him to drive a motor vehicle of a class included in the category or sub-category in respect of which the test is to be taken and a counterpart thereof,

   (ii) except where he has produced an appropriate licence containing his photograph, satisfy the person conducting the test as to his identity in accordance with paragraph (6), and

   (iii) sign a record of his attendence at the test;

   (b) during the test comply with all reasonable instructions given by the invigilator for the purpose of ensuring the proper and orderly conduct of the test.

(4) A person submitting himself for a practical test on a motor vehicle of a class included in any category shall, except in a case where—

   (a) he is exempt from the requirement to pass a theory test by virtue of regulation 42, or

   (b) by virtue of regulation 40(3), no theory test is prescribed for that class,

produce to the person conducting the test before the test commences a valid theory test pass certificate showing that he has passed the theory test prescribed in respect of the same category or a valid certificate corresponding to such a certificate furnished under the law of Northern Ireland.

(5) A person submitting himself for a practical or unitary test shall, before the test commences—

   (a) produce to the person conducting the test an appropriate licence authorising him to drive a motor vehicle of the class on which the test is to be taken and a counterpart thereof,

   (b) except where he has produced an appropriate licence containing his photograph, satisfy the person conducting the test as to his identity in accordance with paragraph (6), and

   (c) sign, on the Driving Test Report Form produced to him by the person conducting the test, a declaration to the effect that there is in force, in relation to the use of the vehicle provided for the purposes of the test, a policy of insurance which complies with the requirements of Part VI of the Traffic Act.

(6) For the purposes of this regulation, a person conducting a test may be satisfied as to a person's identity—

   (a) from a document produced to him which is a document listed in Schedule 6 or is a document of a like nature, or

   (b) if that person's identity is clearly apparent from facts known to, or other evidence in the possession of, the person conducting the test.

(7) In the case of an applicant who is a full-time member of the armed forces of the Crown (to whom the provisions of regulation 11(1) do not apply), he shall before the commencement of a theory test or, as the case may be, a practical or unitary test prescribed in respect of a category specified in column (1) of the table at the end of regulation 11 satisfy the examiner that he has passed the test prescribed in respect of the category specified in column (2) of the table in relation to the first category.

(8) A person submitting himself for a practical test for a licence authorising the driving of a motor vehicle of a class included in category A or P shall before the test commences, unless he is exempt from the requirement imposed by section 89(2A) of the Traffic Act, produce to the examiner a valid certificate furnished under regulation 68(1).

(9) A person submitting himself for a practical or a unitary test shall, during the test—

   (a) except where the test is for a licence authorising him to drive a motor vehicle of a class included in category A, G, H, K or P or a motor vehicle in sub-category B1 which has no seat other than the driver's seat, allow to travel in the vehicle—

      (i) the person authorised to conduct the test; and

      (ii) any person authorised by the Secretary of State to attend the test for the purpose of supervising it or otherwise;

   (b) where the test is for a licence authorising him to drive a motor vehicle of a class included in category A, G, H or P or a motor vehicle in sub-category B1 which has no seat other than the driver's seat, allow the attendance of—

      (i) the person authorised to conduct the test; and

(ii) any person authorised by the Secretary of State for the purpose of supervising the test or otherwise.

(10) In this regulation and regulation 39—

"appropriate licence" means a licence, other than an excepted provisional licence, which—

(a) is valid at the date of the test,

(b) bears the signature of the person who has submitted himself for the test, and

(c) is either—

    (i) a provisional licence authorising the person submitting himself for the test to drive motor vehicles of the same class as the vehicle which he has provided for the test, or

    (ii) a full licence which by virtue of section 98 of the Traffic Act and regulation 19, authorises that person to drive motor vehicles of that class subject to the same conditions as if he were so authorised by a provisional licence, or

    (iii) a Northern Ireland licence corresponding to either of those licences, or

    (iv) a Community licence which, by virtue of section 99A of the Traffic Act and regulation 19, authorises that person to drive motor vehicles of that class subject to the same conditions as if he were so authorised by a provisional licence;

"excepted provisional licence" means a licence which—

(a) was in force at a time before 1st January 1997, and

(b) is issued as a provisional licence in respect of motor vehicles of a class included (by virtue of these Regulations) in category C + E or D + E or sub-categories C1 and D1 (not for hire or reward),

but does not include a licence which was granted to a full-time member of the armed forces of the Crown to whom the provisions of regulation 11(1) do not apply by virtue of paragraph (2) of that regulation.

Examiner's right to refuse to conduct test

39.—(1) Subject to paragraphs (2) and (3), where a person submitting himself for—

(a) a theory test fails to satisfy the person authorised to conduct it that he has complied with any requirement imposed by regulation 38(3), or

(b) a practical or unitary test fails to satisfy the person authorised to conduct it that he has complied with any requirement imposed by paragraphs (4) to (10) of regulation 37 or by paragraph (4), (5), (7) or (8) of regulation 38,

the person authorised to conduct the test must refuse to do so.

(2) Where a person—

(a) fails to produce an appropriate licence as required under paragraph (3)(a)(i) or (5)(a) of regulation 38, or

(b) where he has submitted himself for a practical or unitary test, fails to produce a document required to be produced under paragraph (4) or (8) of that regulation,

if the person authorised to conduct the test—

    (i) is satisfied from other evidence that the document in question exists, and

    (ii) in the case of a person who has failed to produce a licence, is satisfied that the requirements of regulation 38(3)(a)(ii) or (5)(b) have been complied with,

he may conduct the test.

(3) Where a person with special needs has failed to give to the person conducting a theory test such notice of those needs (being not less than 15 working days) as he may reasonably require the person authorised to conduct the test may refuse to do so.

(4) Where a person who requires the assistance of an interpreter at the theory test attends at the test with an interpreter who—

(a) is not approved by the Secretary of State to act as such, or

(b) appears to be acquainted with the test candidate,

the person authorised to conduct the test must refuse to do so.

(5) In this regulation, "special needs" means a reasonable requirement for special treatment during the test arising by virtue of—

(a) the test not being available in a language which the test candidate understands,

(b) the test candidate having reading difficulties, or

(c) the test candidate being physically disabled.

Nature and conduct of tests

Nature of tests other than extended tests

40.—(1) This regulation applies to tests other than extended driving tests.

(2) Subject to the following provisions of this regulation and regulation 42, the test for a licence authorising the driving of a motor vehicle of a class included in category A, B, C, D, or P shall be conducted in two parts, namely—

(a) a theoretical test, and

(b) a practical test of driving skills and behaviour,

and a person taking such a test must pass both parts.

(3) The test for a licence authorising the driving of a motor vehicle of a class included in category B + E, C + E and D + E—

(a) in a case where the test is for a licence authorising the driving of vehicles in sub-category C1 + E and the applicant is the holder of a full licence which was in force at a time before 1st January 1997 and authorises the driving of motor vehicles included in sub-category C1 + E (8.25 tonnes) but not the driving of any other vehicles included in category C + E, shall consist of the specified matters prescribed in respect of the theory test for category C and the specified requirements prescribed in respect of practical test for category C + E, and

(b) in any other case, shall consist of a practical test only.

(4) Where a test is required to be conducted in two parts, a person taking the test—

(a) must pass the theory test before he take the practical test, and

(b) shall not be entitled to apply for an appointment (or, as the case may be, be nominated pursuant to regulation 32(4) or 33(4)) for a practical test in respect of a motor vehicle of a class included in any category until he has been furnished with—

(i) a valid theory test pass certificate stating that he has passed the theory test prescribed in respect of that category, or

(ii) a certificate corresponding to such a certificate furnished under the law of Northern Ireland stating that he has during the relevant period passed the theory test in respect of the same category.

(5) A person shall be treated as having passed—

(a) the theory test if he satisfies the person conducting it that he has a knowledge and sound understanding of the specified matters;

(b) the practical test if he satisfies the person conducting it of his ability to drive safely and to comply with the specified requirements.

(6) The test for a licence authorising the driving of a motor vehicle of a class included in category F, G, H or K shall be a unitary test and a person taking such a test shall be treated as having passed it if he satisfies the person conducting it that he is—

(a) generally competent to drive a vehicle of that class without danger to, and with due consideration for, other road users,

(b) fully conversant with the Highway Code, and

(c) able to comply with the specified requirements.

(7) The practical test and the unitary test shall each be conducted so that—

(a) the person taking the test drives, wherever possible, both on roads outside built-up areas and on urban roads, and

(b) the time during which that person is required to drive on roads is—

(i) in the case of a test for a licence authorising the driving of a class of vehicle included in category B + E, C, C + E, D or D + E, not less than 50 minutes;

(ii) in the case of any other test, not less than 30 minutes.

(8) The theory test shall—

    (a) be conducted as an approved form of examination consisting of 35 questions, the questions being in either a multiple choice or multiple response form and testing a candidate on the specified matters in accordance with Schedule 7;

    (b) have a duration of 40 minutes or, in the circumstances specified in paragraph (9), 80 minutes;

and an approved form of examination is one which is conducted in writing or by means of data recorded on equipment operating automatically in response to instructions given by the candidate.

(9) The circumstances referred to in paragraph (8) are that the candidate requires the assistance of a suitably qualified person at the test by virtue of having reading difficulties.

(10) The specified matters for a theory test for a licence authorising the driving of a motor vehicle of a class included in a category shown in column (1) of the table at the end of this regulation are the matters specified in relation to that category in column (2) of the table.

(11) The specified requirements for a practical or unitary test for a licence authorising the driving of a motor vehicle of a class included in a category shown in column (1) of the table are the requirements specified in relation to that category in column (3) of the table.

TABLE

(1) Category	(2) Specified matters	(3) Specified requirements
A	Matters specified in Part 1 of Schedule 7.	Requirements specified in Part 1 of Schedule 8.
B	Matters specified in Part 2 of Schedule 7.	Requirements specified in Part 2 of Schedule 8.
B+E	—	Requirements specified in Part 2 of Schedule 8.
C	Matters specified in Part 3 of Schedule 7.	Requirements specified in Part 3 of Schedule 8.
D	Matters specified in Part 4 of Schedule 7.	Requirements specified in Part 4 of Schedule 8.
C+E	—	Requirements specified in Part 3 of Schedule 8.
D+E	—	Requirements specified in Part 4 of Schedule 8.
F	—	Requirements specified in Parts 5 and 6 of Schedule 8.
G	—	Requirements specified in Parts 5 and 6 of Schedule 8.
H	—	Requirements specified in Parts 5 and 7 of Schedule 8.
K	—	Requirements specified in Part 5 of Schedule 8.
P	Matters specified in Part 1 of Schedule 7.	Requirements specified in Part 1 of Schedule 8.

Nature of extended driving tests

41.—(1) Where a person is disqualified by order of a court under section 36 of the Offenders Act until he passes an extended driving test, the test which he must pass is a test conducted in accordance with paragraphs (2) to (11) of regulation 40 as modified by virtue of paragraph (2) of this regulation.

(2) For the purpose of an extended driving test, the provisions of regulation 40 shall apply but as if paragraph (1) were omitted and for paragraph (7)(b) there were substituted—

    "(b) the time during which that person is required to drive on roads is not less than 60 minutes".

Exemption from theory test

42.—(1) A person is exempt from the requirement to pass a theory test for the purpose of obtaining a licence authorising him to drive a motor vehicle of a class included in category A if—

(a) he has, on or after 1st July 1996, passed the test prescribed in respect of category P and holds a full licence authorising the driving of a class of vehicles in that category; or

(b) he holds a full licence authorising the driving of motor vehicles either of another class included in category A or of a class included in category B; or

(c) he has passed a Northern Ireland test of competence corresponding to the test mentioned in sub-paragraph (a), or is the holder of a Northern Ireland licence corresponding to a licence mentioned in sub-paragraph (b); or

(d) he has passed a test for a licence authorising the driving of motor vehicles either of another class included in category A or of a class included in category B and is in either case a full-time member of the armed forces of the Crown.

(2) A person is exempt from the requirement to pass a theory test for the purpose of obtaining a licence authorising him to drive a motor vehicle of a class included in category B if—

(a) he has, on or after 1st July 1996, passed the test prescribed in respect of category P and holds a full licence authorising the driving of a class of vehicles in that category; or

(b) he holds a full licence authorising the driving of motor vehicles either of another class included in category B or of a class included in category A; or

(c) he has passed a Northern Ireland test of competence corresponding to the test mentioned in sub-paragraph (a) or is the holder of a Northern Ireland licence corresponding to the licence mentioned in sub-paragraph (b); or

(d) he has passed a test for a licence authorising the driving of motor vehicles either of another class included in category B or of a class included in category A and is in either case a full-time member of the armed forces of the Crown.

(3) A person is exempt from the requirement to pass a theory test for the purpose of obtaining a licence authorising him to drive a motor vehicle of a class included in category C if—

(a) he holds a full licence authorising the driving of motor vehicles of another class included in category C, other than a licence authorising the driving only of vehicles of a class included in sub-category C1 which was in force at a time before 1st January 1997, or a Northern Ireland licence corresponding to such a licence; or

(b) on or after 1st January 1997, he has passed a test for a licence authorising the driving of motor vehicles of another class included in category C and is a full-time member of the armed forces of the Crown.

(4) A person is exempt from the requirement to pass a theory test for the purpose of obtaining a licence authorising him to drive a motor vehicle of a class included in category D if—

(a) he holds a full licence authorising the driving of motor vehicles of another class included in category D other than—

(i) vehicles of a class included in sub-category D1 (not for hire or reward), and

(ii) vehicles in category D which are driven otherwise than for hire or reward;

or a Northern Ireland licence corresponding to such a licence; or

(b) on or after 1st January 1997, he has passed a test prescribed in respect of motor vehicles of another class included in category D and is a full-time member of the armed forces of the Crown.

(5) Where a person is disqualified by order of a court under section 36 of the Offenders Act until he passes the appropriate driving test, he shall not be exempt from the requirement to pass a theory test in respect of any class of motor vehicle by virtue of the foregoing provisions of this regulation until the disqualification is deemed to have expired in relation to that class.

38

(6) Where the Secretary of State has revoked a person's licence or test pass certificate under section 3(2) of, or Schedule 1 to, the Road Traffic (New Drivers) Act 1995 he shall not be exempt from the requirement to pass a theory test in respect of any class of motor vehicle by virtue of the foregoing provisions of this regulation until the day following the date on which he passes a relevant driving test within the meaning of section 4(2) of, or paragraph 6 or 9 of Schedule 1 to, that Act.

Entitlements upon passing test

Entitlement upon passing a test other than an appropriate driving test

43.—(1) Where a person passes a test other than an appropriate driving test prescribed in respect of any category for a licence which (by virtue of regulation 37) authorises the driving of motor vehicles included in that category or in a sub-category thereof, or has passed a Northern Ireland test of competence corresponding to that test, the Secretary of State shall grant to him a licence in accordance with paragraphs (2), (3) and (4).

(2) Subject to regulation 44, the licence shall authorise the driving of all classes of motor vehicle included in that category or sub-category unless—

    (a) the test or, as the case may be, the practical test is passed on a motor vehicle with automatic transmission, in which case it shall authorise the driving only of such classes of vehicle included in that category or sub-category as have automatic transmission;

    (b) the test or, as the case may be, the practical test, is passed on a motor vehicle which is adapted on account of a disability of the person taking the test, in which case it shall authorise the driving only of such classes of vehicle included in that category or sub-category as are so adapted (and for the purposes of this paragraph, a motor bicycle with a side-car may be treated in an appropriate case as a motor vehicle adapted on account of a disability).

(3) The licence shall in addition authorise the driving of all classes of motor vehicle included in a category or sub-category which is specified in column (3) of Schedule 2 as an additional category or sub-category in relation to a category or sub-category specified in column (1) of that Schedule unless—

    (a) the test or, as the case may be, the practical test is passed on a motor vehicle with automatic transmission, in which case it shall (subject to paragraph (4)) authorise the driving only of such classes of vehicle included in the additional category or sub-category as have automatic transmission;

    (b) the test or, as the case may be, the practical test is passed on a motor vehicle which is adapted on account of a disability of the person taking the test in which case it shall authorise the driving only of such classes of vehicle included in the additional category or sub-category as are so adapted.

(4) Where the additional category is F, K or P, paragraph (3)(a) shall not apply.

Entitlement upon passing a test other than an appropriate driving test: category A

44.—(1) This regulation applies where a person has passed a test (or a Northern Ireland test of competence corresponding to such a test) for a licence authorising the driving of motor bicycles of any class other than a class included in sub-category A1.

(2) Where this regulation applies the Secretary of State shall grant to the person who passed the test—

    (a) in a case where he has passed the practical test (or the Northern Ireland test of competence corresponding to the practical test) on a motor bicycle without a side-car, the engine of which has a maximum net power output of not less than 35 kilowatts, a licence authorising him to drive all classes of motor vehicle included in category A;

    (b) subject to paragraph (3), in a case where the practical test (or the Northern Ireland test of competence corresponding to the practical test) was passed on any other motor bicycle without a side-car, a licence authorising him to drive standard motor bicycles;

    (c) in a case where he has passed the practical test (or the Northern Ireland test of competence corresponding to the practical test) on a motor bicycle and side-car combination and the engine of the bicycle has a maximum net power output of not less than 35 kilowatts, a licence authorising him to drive all classes of motor bicycle and side-car combinations included in category A;

(d) subject to paragraph (4), in a case where the practical test (or the Northern Ireland test of competence corresponding to the practical test) was passed on a motor bicycle and a side-car combination the power to weight ratio of which does not exceed 0.16 kw/kg. but which does not fall within paragraph (c), a licence authorising him to drive standard motor bicycles and side-car combinations.

(3) A licence granted to a person by virtue of paragraph (2)(b) shall authorise him to drive all classes of motor vehicle included in category A upon the expiration of the standard access period.

(4) A licence granted to a person by virtue of paragraph (2)(d) shall authorise him to drive all classes of motor bicycle and side-car combinations included in category A upon the expiration of the standard access period.

Upgrading of entitlements by virtue of passing second test

45.—(1) A person who has passed tests for a licence authorising the driving of motor vehicles included in—

(a) category D or sub-category D1 as specified in column (1) of Table A in Schedule 9, and

(b) category C + E or sub-category C1 + E as respectively specified at the top of columns (2) and (3) of Table A,

is deemed, subject to the following paragraphs of this regulation, competent to drive (in addition to the classes of motor vehicle in respect of which the tests were passed) vehicles included in the category or sub-category shown in column (2) or (3) of Table A in relation to the relevant test pass in column (1).

(2) Where, in a case to which paragraph (1) applies, each practical test is passed on a vehicle having automatic transmission the person passing the tests is deemed competent to drive only such classes of vehicle in the upgrade category as have automatic transmission.

(3) A person who has passed a test for a licence authorising the driving of—

(a) motor vehicles included in a category or sub-category specified in column (A) of Table B in Schedule 9 which have automatic transmission, and

(b) motor vehicles included in a category or sub-category specified at the head of one of the columns in that table numbered (1) to (8) which have manual transmission,

is, subject to the following paragraphs of this regulation, deemed competent to drive in addition to the classes of vehicle in respect of which the tests were passed all vehicles included in the category or sub-category shown in the relevant numbered column of Table B in relation to the relevant test pass mentioned in column (A).

(4) Where a person has passed tests for a licence authorising the driving of—

(a) motor vehicles in category D not more than 5.5 metres in length having automatic transmission, and

(b) motor vehicles in category C, other than vehicles in sub-category C1, having manual transmission,

he is deemed competent to drive vehicles in category D not more than 5.5 metres in length which have manual transmission.

(5) In the case of a person who holds a licence which, by virtue of regulation 76 (notwithstanding that he may not have passed a test authorising the driving of such vehicles), authorises the driving of a class of vehicles in category D when used under a section 19 permit or (if not so used) are driven otherwise than for hire or reward, Tables A and B shall be read as if—

(a) for "D" there were substituted "vehicles in category D, driven otherwise than for hire or reward", and

(b) for "D + E" there were substituted "vehicles in category D + E driven otherwise than for hire or reward".

(6) In the case of a person who has passed a test for a licence authorising the driving only of those classes of vehicle in category C + E which are drawbar trailer combinations, paragraphs (1), (2) and (3) and Tables A and B in Schedule 9 shall apply as if he had passed a test for a licence authorising only the driving of the corresponding classes of vehicle in category C.

(7) Where, in Table B, the upgrade category is qualified by the expression "(a)", the person is deemed competent to drive only such classes of vehicle therein as have automatic transmission.

(8) Where a person has passed a test prescribed in respect of category B + E which authorises the driving only of classes of vehicle having automatic transmission and a test prescribed in respect of any class of vehicle in category C or D which authorises the driving of vehicles with manual transmission, he is deemed competent to drive vehicles in category B + E with manual transmission.

(9) Where a person, who is the holder of a licence which authorises the driving of motor vehicles included in categories B and B + E and sub-categories C1, C1 + E (8.25 tonnes), D1 (not for hire or reward) and D1 + E (not for hire or reward) which have automatic transmission, passes a test prescribed in respect of category B, B + E, C or D which authorises the driving of vehicles with manual transmission, he is deemed competent to drive vehicles in category B + E and in sub-categories C1, C1 + E (8.25 tonnes), D1 (not for hire or reward) and D1 + E (not for hire or reward) which have manual transmission.

(10) Where a person has passed tests for a licence authorising the driving of—

   (a) motor vehicles included in category B, other than vehicles included in sub-categories B1 and B1 (invalid carriages), having automatic transmission, and

   (b) motor vehicles included in category B + E, C or D having manual transmission,

he is deemed competent to drive vehicles in category B which have manual transmission.

(11) In this regulation—

   (a) "upgrade category" means the additional category or sub-category which the person passing the tests (or holding the licence and passing the test) is deemed competent to drive by virtue of the relevant provision of this regulation, and

   (b) a reference to a test or a practical test includes, as the case may be, a reference to a Northern Ireland test of competence or a Northern Ireland practical test corresponding thereto.

Entitlement upon passing an appropriate driving test

46.—(1) Where a person—

   (a) is disqualified by order of a court under section 36 of the Offenders Act until he passes the appropriate driving test, and

   (b) passes the appropriate driving test for a licence authorising the driving of a class of motor vehicles included in any category or sub-category,

the disqualification shall, subject to paragraph (8), be deemed to have expired in relation to that class and such other classes of motor vehicle as are specified in paragraphs (2), (3), (4), (5) and (6).

(2) Subject to paragraph (4), the disqualification shall be deemed to have expired in relation to all classes of vehicle included in the category or sub-category referred to in paragraph (1)(b) unless—

   (a) the test or, as the case may be, the practical test is passed on a motor vehicle with automatic transmission, in which case the disqualification shall be deemed to have expired only in relation to such classes of vehicle included in that category or sub-category as have automatic transmission;

   (b) the test or, as the case may be, the practical test is passed on a motor vehicle which is adapted on account of a disability of the person taking the test, in which case the disqualification shall be deemed to have expired only in relation to such classes of motor vehicle included in that category or sub-category as are so adapted (and for the purposes of this paragraph, a motor bicycle with a side-car may be treated in an appropriate case as a motor vehicle adapted on account of a disability).

41

(3) The disqualification shall be deemed to have expired in relation to all classes of vehicle included in any other category which is specified in column (3) of Schedule 2 as being an additional category or sub-category in relation to that category or sub-category unless—

    (a) subject to paragraph (5), the test or, as the case may be, the practical test is passed on a vehicle with automatic transmission, in which case the disqualification shall be deemed to have expired only in relation to such classes of motor vehicle included in the additional category or sub-category as have automatic transmission;

    (b) the test or, as the case may be, the practical test, is passed on a vehicle which is adapted on account of a disability of the person taking the test, in which case the disqualification shall be deemed to have expired only in relation to such classes of motor vehicle included in the additional category or sub-category as are so adapted.

(4) Where, at the date on which a person is disqualified—

    (a) he holds a licence which was granted pursuant to regualtion 44(2)(b) or (d), and

    (b) the standard access period has not expired,

the disqualification shall not, by virtue of paragraph (2) or (7), be deemed to have expired—

    (i) in a case to which regulation 44(2)(b) applies, in relation to large motor bicycles, or

    (ii) in a case to which regulation 44(2)(d) applies, in relation to large motor bicycle and side-car combinations,

until the standard access period has expired.

(5) Paragraph (3)(a) shall not apply where the additional category is F, G, H, K, L or P.

(6) Where the person who is disqualified passes the practical test on a vehicle of a class included in category A, other than sub-category A1, the disqualification shall be deemed to have expired additionally in relation to all classes of vehicle included in—

    (a) categories B, B + E, C, C + E, D and D + E, unless that test is passed on a vehicle with automatic transmission, in which case the disqualification shall be deemed to have expired only in relation to such classes of motor vehicle included in those categories as have automatic transmission;

    (b) categories F, G, H and L.

(7) Where the person who is disqualified passes the practical test on a vehicle of a class included in category B, other than a vehicle included in sub-category B1, the disqualification shall be deemed to have expired additionally in relation to all classes of vehicle included in—

    (a) categories A, B + E, C, C + E, D and D + E, unless that test is passed on a vehicle with automatic transmission, in which case the disqualification shall be deemed to have expired only in relation to such classes of motor vehicle included in those categories as have automatic transmission;

    (b) categories G, H and L.

(8) Where a person is, pursuant to regulation 56, disqualified by the Secretary of State until he passes a driving test prescribed in respect of a class of large goods or passenger-carrying vehicle, the disqualification shall not be deemed to have expired in relation to any class of large goods or passenger-carrying vehicle until he passes that test.

Test results

Evidence of result of theory test

47.—(1) For the purpose of ascertaining whether a candidate has demonstrated a knowledge and sound understanding of the specified matters in accordance with these Regulations the person conducting a theory test shall arrange for the test to be marked—

    (a) in the case of a test conducted before 4th January 2000 in respect of which no request has been made for the theory test pass certificate or failure statement to be furnished on the day of the test, as soon as practicable after completion of the test, and

    (b) in any other case, on the day of the test.

(2) A person conducting the theory test shall, upon completion of the marking of the test, furnish—

(a) a person who passes the test with a theory test pass certificate in the form set out in Part 1 of Schedule 10;

(b) a person who fails to pass the test with a failure statement in the form set out in Part 2 of Schedule 10.

(3) Where a person who has conducted a theory test is satisfied that a theory test pass certificate or a failure statement has been furnished in error to a person who took a theory test, he shall, upon receipt of that document from the person who took the test and subject to paragraph (4), furnish that person with a correct certificate or statement, as the case may be.

(4) Where the person who took the test alleges that a failure statement has been furnished in error returns the statement not later than 14 days after it is furnished to him to the person who conducted the test with a request in writing that the test be remarked, the person who conducted the test shall comply with that request for the purpose of ascertaining whether an error has been made but subject thereto he shall not be obliged to remark any test.

(5) A theory test pass certificate furnished in error, or with an error in the particulars required to be specified in it, may not be presented, in support of an application for a licence, as evidence that a person has passed the test mentioned in such certificate.

(6) A theory test pass certificate shall be valid for the purposes of regulation 38(4) for a period commencing on the date on which the test was taken and ending—

(a) two years later, or

(b) on the date on which the person to whom the certificate was given is disqualified by order of a court under section 36 of the Offenders Act until he passes the appropriate driving test,

whichever is the earlier.

(7) A theory test pass certificate is not valid for the purposes of regulation 38(4) if—

(a) it is furnished in error or with an error in the particulars required to be specified in it; or

(b) the person to whom it is furnished is at that time ineligible, by virtue of an enactment contained in the Traffic Act or these Regulations, to take the test to which the certificate relates.

(8) A person authorised to conduct theory tests by virtue of paragraphs (b), (c), (d) or (e) of regulation 23(1) or regulation 23(2)(b) shall issue theory test pass certificates using forms supplied by the Secretary of State who may make a charge of £5 per form.

Evidence of the result of practical or unitary test

48.—(1) A person conducting a practical or unitary test shall upon completion of the test furnish—

(a) a person who passes the test with a test pass certificate in the form set out in Part 1 of Schedule 11;

(b) a person who fails to pass the test with a statement in the form set out in Part 2 of Schedule 11.

(2) A test pass certificate is invalid if—

(a) the person to whom it is issued is at that time ineligible, by virtue of an enactment contained in the Traffic Act or these Regulations, to take the practical test to which the certificate relates;

(b) at the time when it is issued, the theory test pass certificate produced to the person conducting the test in accordance with regulation 38(4) is invalid by virtue of regulation 47(7).

(3) A person authorised to conduct practical or unitary tests by virtue of sub-paragraphs (b), (c), (d), (e) or (f) of regulation 24(1) or regulation 24(2)(b) shall issue test pass certificates using the forms supplied by the Secretary of State who may make a charge—

(a) in the case of forms supplied to a person authorised under regulation 24(1)(b) or (2)(b), of £2.43 per form, and

(b) in the case of forms supplied to any other person, of £15 per form.

43

PART IV

GOODS AND PASSENGER-CARRYING VEHICLES

General

Part III of the Traffic Act: Prescribed classes of goods and passenger-carrying vehicle

49.—(1) All classes of motor vehicle included in categories C, C+E, D and D+E, except vehicles of classes included in sub-categories C1, C1+E (8.25 tonnes) D1 (not for hire or reward) and D1+E (not for hire or reward), are prescribed for the purposes of section 89A(3) of the Traffic Act.

(2) Subject to paragraph (3), all classes of motor vehicle included in categories C, C+E, D and D+E, except vehicles of classes included in sub-categories C1+E (8.25 tonnes), D1 (not for hire or reward) and D1+E (not for hire or reward), are prescribed for the purposes of section 99(1) and (1A) of the Traffic Act.

(3) In the case of a licence in force at a time before 1st January 1997, paragaph (2) above shall apply as if "C1," was inserted after "sub-categories".

(4) All classes of motor vehicle included in categories C, C+E, D and D+E, except vehicles of classes included in sub-categories C1+E (8.25 tonnes), D1 (not for hire or reward) and D1+E (not for hire or reward), are prescribed for the purposes of section 99A(3) and (4) of the Traffic Act.

Part IV of the Traffic Act: prescribed classes of large goods and passenger-carrying vehicle

50.—(1) Part IV of the Traffic Act and regulations 54 to 57 shall not apply to a large goods vehicle—

    (a) of a class included in category F, G or H or sub-category C1+E (8.25 tonnes), or

    (b) which is an exempted goods vehicle or an exempted military vehicle.

(2) Part IV of the Traffic Act and regulations 54 to 57 shall not apply to a passenger-carrying vehicle manufactured more than 30 years before the date when it is driven and not used for hire or reward or for the carriage of more than eight passengers;

(3) Part IV of the Traffic Act and regulations 54 to 57 shall not apply to a passenger-carrying vehicle when it is being driven by a constable for the purpose of removing or avoiding obstruction to other road users or other members of the public, for the purpose of protecting life or property (including the passenger-carrying vehicle and its passengers) or for other similar purposes.

(4) All classes of large goods and passenger-carrying vehicle to which Part IV of the Traffic Act applies are prescribed for the purposes of section 117(7) and 117A(6) of the Traffic Act.

Exempted goods vehicles and military vehicles

51.—(1) For the purposes of this Part of these Regulations, an exempted goods vehicle is a vehicle falling within any of the following classes—

    (a) a goods vehicle propelled by steam;

    (b) any road construction vehicle used or kept on the road solely for the conveyance of built-in road construction machinery (with or without articles or materials used for the purpose of that machinery);

    (c) any engineering plant other than a mobile crane;

    (d) a works truck;

    (e) an industrial tractor;

    (f) an agricultural motor vehicle which is not an agricultural or forestry tractor;

    (g) a digging machine;

    (h) a goods vehicle which, in so far as it is used on a road—

        (i) is used only in passing from land in the occupation of a person keeping the vehicle to other land in the occupation of that person, and

     (ii) is not used on roads for distances exceeding an aggregate of 9.7 kilometres in any calendar week;

(j) a goods vehicle, other than an agricultural motor vehicle, which—

     (i) is used only for purposes relating to agriculture, horticulture or forestry,

     (ii) is used on roads only in passing between different areas of land occupied by the same person, and

     (iii) in passing between any two such areas does not travel a distance exceeding 1.5 kilometres on roads;

(k) a goods vehicle used for no other purpose than the haulage of lifeboats and the conveyance of the necessary gear of the lifeboats which are being hauled;

(l) a goods vehicle manufacturered before 1st January 1960, used unladen and not drawing a laden trailer;

(m) an articulated goods vehicle the unladen weight of which does not exceed 3.05 tonnes;

(n) a goods vehicle in the service of a visiting force or headquarters as defined in the Visiting Forces and International Headquarters (Application of Law) Order 1965(**a**);

(o) a goods vehicle driven by a constable for the purpose of removing or avoiding obstruction to other road users or other members of the public, for the purpose of protecting life or property (including the vehicle and its load) or for other similar purposes;

(p) a goods vehicle fitted with apparatus designed for raising a disabled vehicle partly from the ground and for drawing a disabled vehicle when so raised (whether by partial superimposition or otherwise) being a vehicle which—

     (i) is used solely for dealing with disabled vehicles;

     (ii) is not used for the conveyance of any goods other than a disabled vehicle when so raised and water, fuel, accumulators and articles required for the operation of, or in connection with, such apparatus or otherwise for dealing with disabled vehicles; and

     (iii) has an unladen weight not exceeding 3.05 tonnes;

(q) a passenger-carrying vehicle recovery vehicle; and

(r) a mobile project vehicle.

(2) For the purposes of this Part of these Regulations, an exempted military vehicle is a large goods or passenger-carrying vehicle falling withing any of the following classes—

(a) a vehicle designed for fire fighting or fire salvage purposes which is the property of, or for the time being under the control of, the Secretary of State for Defence, when being driven by a member of the armed forces of the Crown;

(b) a vehicle being driven by a member of the armed forces of the Crown in the course of urgent work of national importance in accordance with an order of the Defence Council in pursuance of the Defence (Armed Forces) Regulations 1939(**b**) which were continued permanently in force, in the form set out in Part C of Schedule 2 to the Emergency Laws (Repeal) Act, 1959(**c**), by section 2 of the Emergency Powers Act 1964(**d**); or

(c) an armoured vehicle other than a track-laying vehicle which is the property of, or for the time being under the control of, the Secretary of State for Defence.

(3) In this Regulation—

"digging machine" has the same meaning as in paragraph 4(4) of Schedule 1 to the Vehicle Excise and Registration Act 1994;

"agricultural motor vehicle", "engineering plant", "industrial tractor" and "works truck" have the same meaning as in regulation 3(2) of the Construction and Use Regulations;

"public road" has the same meaning as in section 62(1) of the Vehicle Excise and Registration Act 1994;

"road construction machinery" means a machine or device suitable for use for the construction and repair of roads and used for no purpose other than the construction and repair of roads; and

---

(**a**) S.I. 1965/1536.
(**b**) S.R. & O. 1939/1304.
(**c**) 1959 c. 19.
(**d**) 1964 c. 38.

"road construction vehicle" means a vehicle which—

(a) is constructed or adapted for use for the conveyance of road construction machinery which is built in as part of, or permanently attached to, that vehicle, and

(b) is not constructed or adapted for the conveyance of any other load except articles and materials used for the purposes of such machinery.

Correspondences

52.—(1) For the purposes of section 89A(5) of the Traffic Act, a heavy goods vehicle or public service vehicle of a class specified in column (1) of the table at the end of this regulation corresponds to a class of large goods vehicle or passenger-carrying vehicle, as the case may be, specified in column (2) of that table in relation to the class of vehicle in column (1).

(2) For the purposes of paragraph (1), where a heavy goods vehicle driver's licence held before 1st April 1991 was restricted to vehicles having a permissible maximum weight not exceeding 10 tonnes by virtue of—

(a) paragraph 3(3) and (5) of Schedule 2 to the Road Traffic (Drivers' Ages and Hours of Work) Act 1976(**a**); or

(b) paragraph (1) or (2) of regulation 31 of the Heavy Goods Vehicles (Drivers' Licences) Regulations 1977(**b**);

before those enactments ceased to have effect, such restriction shall be disregarded.

TABLE

(1) Class of heavy goods or public service vehicle	(2) Corresponding class of large goods or passenger-carrying vehicle
Heavy goods vehicles	*Large goods vehicles*
1	Categories C and C + E
1A	Categories C and C + E (limited, in each case, to vehicles with automatic transmission)
2	Category C and vehicles in category C + E which are drawbar trailer combinations
2A	Category C and vehicles in category C + E which are drawbar trailer combinations (limited, in each case, to vehicles with automatic transmission)
3	Category C and vehicles in category C + E which are drawbar trailer combinations
3A	Category C and vehicles in category C + E which are drawbar trailer combinations (limited, in each case, to vehicles with automatic transmission)
Public Service Vehicles	*Passenger-carrying vehicles*
1	Categories D and D + E
1A	Categories D and D + E (limited, in each case, to vehicles with automatic transmission)
2	Categories D and D + E
2A	Categories D and D + E (limited, in each case, to vehicles with automatic transmission)
3	Category D
3A	Category D (limited to vehicles with automatic transmission)
4	Sub-category D1 and vehicles in category D not more than 5.5 metres in length
4A	Sub-category D1 and vehicles in category D not more than 5.5 metres in length (limited, in each case, to vehicles with automatic transmission)

(**a**) 1976 c. 3.
(**b**) S.I. 1977/1309, to which there were amendments not relevant to these Regulations.

Part IV of the Traffic Act: dual purpose vehicles

53.—(1) Except in the case of a vehicle mentioned in paragraph (2), Part IV of the Traffic Act and regulations 54 to 57 shall apply to dual purpose vehicles to the extent that they apply to passenger-carrying vehicles.

(2) Part IV of the Traffic Act and regulations 54 to 57 shall apply to any dual purpose vehicle which is—

(a) driven by a member of the armed forces of the Crown, and

(b) used to carry passengers for naval, military or air force purposes,

to the extent that they apply to large goods vehicles.

Persons under the age of 21

Large goods vehicles drivers' licences issued to persons under the age of 21: trainee drivers

54.—(1) A large goods vehicle driver's licence granted to a person under the age of 21 is subject to the conditions prescribed, for the purposes of section 114(1) of the Traffic Act, in the following paragraphs.

(2) In the case of an LGV trainee driver's licence, whether issued as a provisional or a full licence or treated as a provisional licence by virtue of section 98 of the Traffic Act and regulation 19, the holder shall not drive a large goods vehicle of any class which the licence authorises him to drive unless—

(a) he is a registered employee of a registered employer, and

(b) the vehicle is a large goods vehicle of a class to which his training agreement applies and is owned or operated by that registered employer or by a registered LGV driver training establishment.

(3) In the case of a licence held by a person who is a member of the armed forces of the Crown, the holder shall not drive a large goods vehicle of any class unless it is owned or operated by the Secretary of State for Defence and is being used for naval, military or air force purposes.

(4) In the case of an LGV trainee driver's full licence, the holder shall not drive a large goods vehicle of any class if the vehicle is being used to draw a trailer except under the supervision of a person who is present with him in the vehicle and who holds a full large goods vehicle driver's licence authorising the driving of a vehicle of that class which is not an LGV trainee driver's licence.

(5) In the case of an LGV trainee driver's full licence authorising the driving of a class of vehicles included in category C, the holder shall not drive large goods vehicles of a class included in category C + E, other than vehicles included in sub-category C1 + E the maximum authorised mass of which does not exceed 7.5 tonnes, as if he were authorised by a provisional licence to do so before the expiration of a period of two years commencing on the date on which he passed the test for that full licence.

(6) In this regulation—

"LGV trainee driver's licence" means a large goods vehicle driver's licence which—

(a) authorises its holder to drive vehicles of a class included in category C or C + E,

(b) is held by a person, other than a member of the armed forces of the Crown, who was under the age of 21 on the date of the application, and

(c) is in force for a period during the whole or part of which that person is under the age of 21;

"registered", in relation to an employee, employer or training establishment, means registered for the time being by the Training Committee in accordance with the Training Scheme;

"training agreement", in relation to an individual who is undergoing, or is to undergo, driver training under the Training Scheme, means the agreement between that individual and a registered employer;

"the Training Committee" means the National Joint Training Committee for Young LGV Drivers in the Road Goods Transport Industry which is referred to in the Training Scheme;

"the Training Scheme" means the Young Large Goods Vehicle (LGV) Drivers Training Scheme which has been established by the Road Haulage and Distribution Training Council and approved by the Secretary of State for the purpose of regulations under section 101(2) of the Traffic Act on 30th September 1996 for training young drivers of large goods vehicles.

Drivers' conduct

Large goods vehicle drivers' licences and LGV Community licences: obligatory revocation or withdrawal and disqualification

55.—(1) The prescribed circumstances for the purposes of section 115(1)(a) of the Traffic Act are that, in the case of the holder of a large goods vehicle driver's licence who is under the age of 21, he has been convicted (or is, by virtue of section 58 of the Offenders Act, to be treated as if he had been convicted) of an offence as a result of which the number of penalty points to be taken into account under section 29 of the Offenders Act(**a**) exceeds three.

(2) The prescribed circumstances for the purposes of section 115A(1)(a) of the Traffic Act are that, in the case of the holder of an LGV Community licence who is under the age of 21, he has been convicted (or is, by virtue of section 58 of the Offenders Act, to be treated as if he had been convicted) of an offence as a result of which the number of penalty points to be taken into account under section 29 of the Offenders Act exceeds three.

(3) Where—

(a) a large goods vehicle drivers' licence is revoked under section 115(1)(a) of the Traffic Act, or

(b) the Secretary of State serves a notice on a person in pursuance of section 115A(1)(a) of that Act,

the cases in which the person whose licence has been revoked or, as the case may be, on whom the notice has been served must be disqualified indefinitely or for a fixed period shall be determined by the Secretary of State.

(4) Where the Secretary of State makes a determination under paragraph (3) that a person is to be disqualified for a fixed period he shall be disqualified until he reaches 21 years of age or for such longer period as the Secretary of State shall determine.

Holders of licences who are disqualified by order of a court

56.—(1) This regulation applies where a person's large goods vehicle or passenger-carrying vehicle driver's licence is treated as revoked by virtue of section 37(1) of the Offenders Act (effect of disqualification by court order) and where it applies subsections (1) and (2) of section 117 of the Traffic Act are modified in accordance with paragraphs (2) to (6).

(2) Where the licence which is treated as revoked is a large goods vehicle driver's licence held by a person under the age of 21—

(a) the Secretary of State must order that person to be disqualified either indefinitely or for a fixed period, and

(b) where the Secretary of State determines that he shall be disqualified for a fixed period, he must be disqualified until he reaches the age of 21 or for such longer period as the Secretary of State determines.

(**a**) Section 29 was inserted by section 28 of the Road Traffic Act 1991.

(3) Where the licence which is treated as revoked is a large goods vehicle driver's licence held by any other person or is a passenger-carrying vehicle driver's licence—

    (a) the Secretary of State may order that person to be disqualified either indefinitely or for such fixed period as he thinks fit, or

    (b) except where the licence is a provisional licence, if it appears to the Secretary of State that, owing to that person's conduct, it is expedient to require him to comply with the prescribed conditions applicable to provisional licences until he passes a test, the Secretary of State may order him to be disqualified for holding or obtaining a full licence until he passes a test.

(4) Where the Secretary of State orders him to be disqualified until he passes a test, that test shall be a test prescribed by these Regulations for a licence authorising the driving of any class of vehicle in category C (other than sub-category C1), C + E, D or D + E which, prior to his disqualification by order of the court, he was authorised to drive by the revoked licence.

(5) Any question as to whether a person—

    (a) shall be disqualified indefinitely or for a fixed period or until he passes a test, or

    (b) if he is to be disqualified for a fixed period, what that period should be, or

    (c) if he is to be disqualified until he passes a test, which test he should be required to pass,

may be referred by the Secretary of State to the traffic commissioner.

(6) Where the Secretary of State determines that a person shall be disqualified for a fixed period, that period shall commence on the expiration of the period of disqualification ordered by the court.

(7) Where this regulation applies, subsections (3) to (6) of section 116 of the Traffic Act shall apply, but as if—

    (a) subsection (4)(a) were omitted,

    (b) for the words "in any other case, revoke the licence or suspend it" in subsection (4)(b) there were substituted "suspend the licence", and

    (c) the references to sections 115(1) and 116(1) of that Act were references to this regulation.

Removal of disqualification

57.—(1) Subject to paragraphs (2) and (3), the Secretary of State may remove a disqualification for a period of more than two years imposed under section 117(2)(a) of the Traffic Act, after consultation with the traffic commissioner in a case which was referred to him, if an application for the removal of the disqualification is made after the expiration of whichever is relevant of the following periods commencing on the date of the disqualification—

    (a) two years, if the disqualification is for less than four years;

    (b) one half of the period of the disqualification, if it is for less than ten years, but not less than four years;

    (c) five years in any other case.

(2) An application may not be made if the applicant has during the relevant period been convicted (or treated as convicted) of an offence by virtue of which he has incurred—

    (a) penalty points, or

    (b) an endorsement of a Northern Ireland driving licence held by him, or of its counterpart, with particulars of a conviction pursuant to provisions for the time being in force in Northern Ireland that correspond to sections 44 and 45 of the Offenders Act.

(3) Where an application under paragraph (1) for the removal of a disqualification is refused, a further such application shall not be entertained if made within three months after the date of refusal.

PART V

APPROVED TRAINING COURSES FOR RIDERS OF MOTOR BICYCLES AND MOPEDS

Approved training courses

Provision of approved training courses

58.—(1) For the purposes of section 97(3)(e) of the Traffic Act an approved training course is a course for riders of motor bicycles or mopeds both complying with and conducted in accordance with this Part of these Regulations and approved by the Secretary of State.

(2) An approved training course may be provided—

(a) by the Secretary of State, in so far as concerns the instruction of persons in the civil service of the Crown under his department,

(b) by the Secretary of State for Defence, in so far as concerns the instruction of persons in the service of the Crown under his department, and

(c) by any chief officer of police, in so far as concerns the instruction of—

(i) members of the police force of which he is the chief officer, or

(ii) persons employed in the driving of motor vehicles for police purposes by the police authority for the area in respect of which he is the chief officer or by the Receiver for the Metropolitan Police District,

if that person satisfies the conditions mentioned in paragraph (4).

(3) A person may apply to the Secretary of State to be authorised to provide approved training courses and the Secretary of State may give such authorisation subject to any conditions which he thinks fit to impose if he is satisfied that the applicant satisfies the conditions mentioned in paragraph (4).

(4) The conditions specified in paragraphs (2) and (3) are that he—

(a) is a fit and proper person to conduct courses,

(b) will make proper arrangements for the conduct of courses in accordance with these Regulations, and

(c) will keep proper records of courses and the results thereof.

(5) In this Part of these Regulations—

"approved training body" means a person authorised to provide approved training courses under this Part;

"approved training course" has the meaning given in paragraph (1);

"prescribed training course" means a course containing the elements prescribed under the regulation 59.

Nature and conduct of training courses

59.—(1) A training course for riders of motor bicycles and mopeds may not be approved by the Secretary of State unless it comprises elements (A) to (E) set out in Schedule 12.

(2) Before any practical instruction is given to riders on an approved training course all the requirements of element (A) of the course must be fulfilled.

(3) To complete an approved training course successfully, a rider of a motor bicycle or moped must satisfy the approved training body or a certified instructor acting on his behalf as to each of the following matters in the following sequence—

(a) that he has fulfilled the requirements set out in element (B) of the course; and

(b) that he can execute the manoeuvres set out in element (C) of the course; and

(c) that all the requirements of element (D) of the course have been fulfilled; and

(d) that he rides safely on roads in a variety of road traffic situations, including as many as practicable of those set out in element (E) of the course.

Instructors

Certified Instructors

60.—(1) No person may conduct instruction in the riding of motor bicycles or mopeds as part of an approved training course except in accordance with this regulation and regulations 61 to 68.

(2) Subject to the following provisions of this regulation, an approved training body may authorise persons to conduct on his behalf instruction of persons in the riding of learner motor bicycles and mopeds.

(3) A person may not be authorised under paragraph (2) unless at the date of authorisation he satisfies the following conditions, namely that—

    (a) he is a fit and proper person to be an instructor;

    (b) he holds a full licence authorising the driving of vehicles in category A other than vehicles included in sub-category A1;

    (c) either—

        (i) in the case of a person who was authorised to conduct instruction by an approved training body in accordance with regulations in force on 30th January 1998, he had held that licence for a period of, or periods amounting in aggregate to, not less than two years, or

        (ii) in any other case, he is at least 21 years of age and has held that licence for a period of, or periods amounting in aggregate to, not less than three years; and

    (d) he has either—

        (i) successfully completed the Secretary of State's assessment course for certified instructors, or

        (ii) been fully trained by a certified instructor who has successfully completed such a course and assessed by that instructor to be capable of acting as a certified instructor.

(4) An authorisation given to a person under paragraph (2) shall be of no effect unless—

    (a) the approved training body has notified the Secretary of State in writing of the proposed authorisation,

    (b) the Secretary of State has approved the authorisation in writing, and

    (c) there is in force in respect of that person a valid certificate, in the form set out in Part 1 of Schedule 13, issued by the Secretary of State to the approved training body giving the authorisation.

(5) A person in respect of whom a certificate issued under paragraph (4)(c) is in force—

    (a) shall be known as a certified instructor, and

    (b) shall be entitled to conduct approved training courses, and

    (c) in the case of a person who has successfully completed the Secretary of State's assessment course for certified instructors, shall be entitled to train other persons and to assess their capability to act as certified instructors.

(6) Where a person who is an approved training body satisfies the conditions set out in paragraph (3), the Secretary of State may issue a certificate in respect of him under paragraph (4)(c) and while that certificate is in force—

    (a) he shall be known as a certified instructor,

    (b) he shall be entitled to conduct approved training courses, and

    (c) in the case of a person who has successfully completed the Secretary of State's assessment course for certified instructors, he shall be entitled to train other persons and to assess their capability to act as certified instructors.

(7) A certificate issued pursuant to paragraph (4)(c) shall be valid for a period of four years but may be renewed upon application being made to the Secretary of State by the approved training body who authorised the instructor.

(8) When conducting an approved training course a certified instructor shall carry with him the certificate issued in respect of him by the Secretary of State and shall, upon being required to do so by a constable or the Secretary of State, produce it for examination.

(9) In this Part of these Regulations "certified instructor" has the meaning given in paragraph (5)(a) or (6)(a) as the case may be.

Persons authorised as assistant instructors

61.—(1) Subject to paragraphs (2) and (3), a person authorised or deemed to be authorised as an assistant instructor by virtue of regulations in force on 30th January 1998 shall be entitled to conduct, on behalf of an approved training body, the instruction of riders of motor bicycles in all elements other than element (E) of the prescribed training course.

(2) No person authorised or deemed to be authorised as an assistant instructor may conduct instruction if at any time he ceases to hold a full licence authorising the driving of vehicles in category A (other than a licence authorising the driving only of vehicles in sub-category A1) or if the Secretary of State, being satisfied that he is not a fit and proper person to conduct instruction, withdraws approval of his authorisation to act as an assistant instructor.

(3) No person shall be entitled to conduct training otherwise than as a certified instructor or certified direct access instructor after 30th January 2002.

Withdrawal of approval to provide training courses or to act as instructor

62.—(1) The Secretary of State may at any time by notice in writing withdraw an authorisation given under regulation 58(3), an approval given under regulation 60(4)(b) or an authorisation granted by virtue of regulation 61(1).

(2) Where the Secretary of State withdraws an authorisation given under regulation 58(3)—
  (a) the approval of that person for the purposes of that regulation, and
  (b) the authority of that person, and of any other person whom he has approved to act as a certified or assistant instructor,
shall cease forthwith and the person whose approval is withdrawn shall, before the expiration of a period of 28 days commencing on the date of withdrawal, return to the Secretary of State all certificates which were issued to him under regulation 60(4)(c) and all forms for certificates which were supplied to him under regulation 68(3).

(3) Where the Secretary of State withdraws an approval given under regulation 60(4)(b) or an authorisation granted by virtue of regulation 61(1)—
  (a) the authority of the person whose approval to act (as the case may be) as a certified or assistant instructor is withdrawn shall cease forthwith, and
  (b) in the case of the withdrawal of an approval given in respect of a certified instructor, the person whose approval is withdrawn shall as soon as is reasonably practicable return the certificate issued under regulation 60(4)(c) and all forms of certificates which were supplied to him under regulation 68(3) to the approved training body who authorised him who must, on receiving the certificate issued under regulation 60(4)(c), return it to the Secretary of State.

Cessation of conduct of training

63.—(1) Where a certified instructor authorised by an approved training body under regulation 60(2) ceases to conduct instruction on behalf of the body who authorised him, he shall as soon as is reasonably practicable return the certificate issued under regulation 60(4)(c) and all forms of certificates which were supplied to him under regulation 68(3) to the approved training body who must, on receiving the certificate issued under regulation 60(4)(c), return it to the Secretary of State.

(2) Where an approved training body who is also entitled under regulation 60(6) to conduct instruction as a certified instructor ceases to conduct such instruction, he shall immediately return the certificate issued under regulation 60(4)(c) and (unless the Secretary of State agrees otherwise) all forms of certificates which were supplied to him under regulation 68(3) to the Secretary of State.

Approved training courses conducted on large motor bicycles

64.—(1) An approved training course for a person holding a provisional licence authorising the driving of large motor bicycles and undertaken by him on a motor bicycle other than a learner motor bicycle must be conducted by a certified direct access instructor.

(2) "Certified direct access instructor" means a person authorised (or deemed to have been authorised) in accordance with regulation 65.

Certified direct access instructors

65.—(1) An approved training body may, subject to the following provisions of this regulation, authorise instructors to conduct on his behalf the instruction of persons who hold provisional licences authorising the riding of large motor bicycles in the riding of motor bicycles other than learner motor bicycles.

(2) A person may not be authorised under paragraph (1) unless he—

(a) holds a full licence to drive motor bicycles,

(b) either—

(i) was authorised on 30th January 1998 to conduct instruction by an approved training body in accordance with these Regulations and has held that licence for a period of, or periods amounting in aggregate to, not less than 2 years, or

(ii) if he was not so authorised, is at least 21 years of age and has held that licence for a period of, or periods amounting in aggregate to, not less than 3 years.

(c) is a certified instructor, and

(d) has successfully completed the Secretary of State's assessment course for certified direct access instructors.

(3) An authorisation given under paragraph (1) shall be of no effect unless—

(a) the person whom the approved training body proposes to authorise, or another person who is at that time validly authorised by the approved training body to provide instruction in the riding of large motor bicycles, has successfully completed the Secretary of State's assessment course for certified instructors in addition to the assessment course for direct access instructors,

(b) the approved training body has notified the Secretary of State in writing of the proposed authorisation, and

(c) the Secretary of State has approved the authorisation in writing.

(4) An authorisation given under paragraph (1) shall be of no effect in the case of a direct access instructor unless there is in force in respect of him a valid certificate, in the form set out in Part 2 of Schedule 13, issued by the Secretary of State to the person who has authorised him under paragraph (1).

(5) The Secretary of State may at any time by notice in writing withdraw an approval given under paragraph (3)(c) and any authorisation given under paragraph (1) shall cease to have effect from the date of such notice.

(6) Any authorisation given under paragraph (1) by an approved training body shall cease to have effect if at any time there ceases to be a person who—

(a) is validly authorised by that approved training body to conduct instruction in accordance with paragraph (1), and

(b) has successfully completed the Secretary of State's assessment course for certified instructors in addition to the assessment course for direct access instructors.

(7) Where—

(a) a person who is an approved training body satisfies the conditions set out in paragraph (2),

(b) either he or another person who is at that time validly authorised by him to provide instruction in the riding of large motor bicycles has successfully completed the Secretary of State's assessment course for certified instructors in addition to the assessment course for direct access instructors, and

(c) there is in force in respect of that training body a valid certificate issued by the Secretary of State under paragraph (4),

he shall be deemed to have been authorised under paragraph (1) as a certified direct access instructor.

(8) Regulations 60(7) and (8), 62(3) and 63 shall apply in respect of a certified direct access instructor as they apply in respect of a certified instructor as if the references therein to the issue of certificates and the giving or withdrawal of approval were references to the issue of certificates and the giving or withdrawal of approval under this regulation.

(9) Where an authorisation given, or deemed to be given, under this regulation in respect of a certified direct access instructor ceases to have effect by virtue of any of the foregoing provisions of this regulation that instructor shall as soon as is reasonably practicable return the certificate issued under paragraph (4) to the approved training body who must, on receiving it, return it immediately to the Secretary of State.

Miscellaneous

Eligibility to undertake approved training course

66. No person shall be eligible to undertake an approved training course unless at the time he undertakes it he holds a provisional licence authorising him to drive a motor bicycle or moped of the class on which the course is to be undertaken or is entitled, by virtue of section 98 or 99A of the Traffic Act and regulation 19, to drive a motor bicycle of that class subject to the same conditions as the holder of a provisional licence.

Ratio of trainees to instructors

67.—(1) Where, during an approved training course, more than one person is receiving on-site instruction and practical on-site riding as part of elements (B) and (C) of the prescribed training course—

(a) in the case of instruction or riding which may under these Regulations be conducted by a certified or an assistant instructor, there shall be no more than four such persons in the charge of any one instructor at any one time,

(b) in the case of instruction or riding which must under regulation 64 be conducted by a certified direct access instructor, there shall be no more than two such persons in the charge of any one instructor at any one time.

(2) Subject to paragraph (3), when riders are undertaking element (E) of the prescribed training course—

(a) there must be no more than two riders in the charge of any one certified or certified direct access instructor at any one time, and

(b) the instructor must be able to communicate with each rider by means of a radio which is not hand-held while in operation.

(3) The requirement specified in paragraph (2)(b) shall not apply in the case of a rider who is unable, by reason of impaired hearing, to receive directions from the instructor by radio where the rider and the instructor are employing a satisfactory means of communication which they have agreed before the start of element (E).

Evidence of successful completion of course

68.—(1) The certified instructor or the certified direct access instructor who conducted element (E) of the prescribed training course shall furnish a person who successfully completes an approved training course with a certificate in the form set out in Part 3 of Schedule 13 and signed by that instructor.

(2) A certificate under paragraph (1) is not valid either for the purposes of section 97(3)(e) of the Traffic Act or as evidence of the succesful completion of an approved training course for the purposes of regulation 38(8)—

(a) if the person to whom it is issued is at the time of issue ineligible to undertake the training course and

(b) after the earlier of the following dates, namely—

    (i) the last day of the period of 3 years commencing with the date of the certificate, or

    (ii) in a case where the person to whom the certificate was given is later disqualified by order of a court under section 36 of the Offenders Act, the date on which the order is made.

(3) A certified instructor or a certified direct access instructor shall issue certificates using forms supplied by the Secretary of State to the approved training body and the Secretary of State may make a charge of £8 per form.

(4) An approved training body may, if satisfied that a certificate issued to a person who has successfully completed an approved training course conducted by that body has been lost or destroyed, issue a duplicate certificate but may not make a charge exceeding £20 in respect of the issue of any one certificate.

Exemptions from Part V

69.—(1) Subject to paragraph (2), section 98(3)(c) of the Traffic Act shall not apply to a person who is a provisional entitlement holder by virtue of having passed a test for the time being prescribed in respect of category P on or after 1st December 1990 and such a person shall be exempt from the requirement imposed by section 89(2A) of that Act.

(2) Paragraph (1) shall cease to apply to a person if he is disqualified by order of a court under section 36 of the Offenders Act.

(3) A provisional licence or provisional entitlement holder who is resident on an exempted island shall be exempt from the requirement imposed by section 89(2A) of the Traffic Act in respect of a test of competence to drive a motor bicycle of any class taken, or to be taken, on an island, whether or not that island is an exempted island.

(4) A provisional licence holder who is resident on an exempted island shall be exempt from the restriction imposed by section 97(3)(e) of the Traffic Act if he satisfies either of the conditions set out in paragraph (6).

(5) Section 98(3)(c) of the Traffic Act shall not apply to a provisional entitlement holder who is resident on an exempted island if he satisfies either of the conditions set in paragraph (6).

(6) The conditions referred to in paragraphs (4) and (5) are that he is—

    (a) driving on an exempted island, whether or not he is also resident on that island; or

    (b) driving on an island which is not an exempted island for the purpose of—

        (i) undertaking, or travelling to or from, an approved training course,

        (ii) undergoing, or travelling to or from a place where he is to take or where he has taken, a test of competence prescribed in respect of category A or P.

(7) In this regulation—

"exempted island" means any island in Great Britain other than—

    (a) the Isle of Wight, the island which comprises Lewis and Harris, the island which comprises North Uist, Benbecula and South Uist, Mainland Orkney and Mainland Shetland, and

    (b) any other island from which motor vehicles not constructed or adapted for special purposes can at some time be conveniently driven to a road in any other part of Great Britain because of the presence of a bridge, tunnel, ford or other way suitable for the passage of such motor vehicles;

"provisional licence holder" means a person who holds a provisional licence which, subject to section 97(3) of the Traffic Act, authorises the driving of motor bicycles of any class; and

"provisional entitlement holder" means a person who holds a full licence which is treated, by virtue of section 98 of the Traffic Act and regulation 19, as authorising him to drive motor bicycles of any class as if he held a provisional licence therefor.

PART VI

DISABILITIES

Licence groups

70.—(1) In this Part of these Regulations—

"Group 1 licence" means a licence in so far as it authorises its holder to drive classes of motor vehicle included in—

    (a) categories A, B, B + E, F, G, H, K, L and P,

    (b) the former category N,

"Group 2 licence" means, subject to paragraphs (2) and (3), a licence in so far as it authorises its holder to drive classes of motor vehicle included in any other category, and

"licence" includes, unless the context otherwise requires, a Northern Ireland licence and a Community licence.

(2) In so far as a licence authorises its holder to drive vehicles of a class included in sub-categories C1, C1 + E (8.25 tonnes), D1 (not for hire or reward) and D1 + E (not for hire or reward) it is a Group 1 licence while it remains in force if—

    (a) it was in force at a time before 1st January 1997, or

    (b) it is granted upon the expiry of a licence which was in force at a time before 1st January 1997 and comes into force not later than 31st December 1997.

(3) Subject to paragraph (6)(d) of regulation 7, a licence shall be a Group 1 licence in so far as it authorises, by virtue of paragraphs (4), (5) and (6) of that regulation, the driving of a class of motor vehicles which is not included in a category or sub-category specified in relation to a Group 1 licence in paragraph (1) or (2) above.

Disabilities prescribed in respect of Group 1 and 2 licences

71.—(1) The following disabilities are prescribed for the purposes of section 92(2) of the Traffic Act as relevant disabilities in relation to an applicant for, or a person who holds, a Group 1 or Group 2 licence—

    (a) epilepsy;

    (b) severe mental disorder;

    (c) liability to sudden attacks of disabling giddiness or fainting which are caused by any disorder or defect of the heart as a result of which the applicant for the licence or, as the case may be, the holder of the licence has a device implanted in his body, being a device which, by operating on the heart so as to regulate its action, is designed to correct the disorder or defect;

    (d) liability to sudden attacks of disabling giddiness or fainting, other than attacks falling within paragraph (1)(c); and

    (e) persistent misuse of drugs or alcohol, whether or not such misuse amounts to dependency.

(2) The disability prescribed in paragraph (1)(c) is prescribed for the purpose of section 92(4)(b) of the Traffic Act in relation to an applicant for a Group 1 or Group 2 licence if the applicant suffering from that disability satisfies the Secretary of State that—

    (a) the driving of a vehicle by him in pursuance of the licence is not likely to be a source of danger to the public; and

    (b) he has made adequate arrangements to receive regular medical supervision by a cardiologist (being a supervision to be continued throughout the period of the licence) and is conforming to those arrangements.

(3) The following disabilities are prescribed for the purposes of paragraphs (a) and (c) of section 92(4) of the Traffic Act namely, any disability consisting solely of any one or more of—

    (a) the absence of one or more limbs,

    (b) the deformity of one or more limbs, or

    (c) the lost of use of one or more limbs, which is not progressive in nature.

(4) In this regulation—

(a) in paragraph (1)(b), the expression "severe mental disorder" includes mental illness, arrested or incomplete development of the mind, psychopathic disorder and severe impairment of intelligence or social functioning;

(b) in paragraph (2)(b), the expression "cardiologist" means a registered medical practitioner who specialises in disorders or defects of the heart and who, in that connection, holds a hospital appointment;

(c) in paragraph (3), references to a limb include references to a part of a limb, and the reference to loss of use, in relation to a limb, includes a reference to a deficiency of limb movement or power.

Disabilities prescribed in respect of Group 1 licences

72.—(1) There is prescribed for the purposes of section 92(2) of the Traffic Act as a relevant disability in relation to an applicant for, or a person who holds, a Group 1 licence, the inability to read in good light (with the aid of corrective lenses if necessary) a registration mark fixed to a motor vehicle and containing letters and figures 79.4 millimetres high at a distance of—

(a) 12.3 metres, in the case of an applicant for a licence authorising only the driving of motor vehicles of a class included in category K, or

(b) 20.5 metres, in any other case.

(2) Epilepsy is prescribed for the purposes of section 92(4)(b) of the Traffic Act in relation to an applicant for a Group 1 licence who either—

(a) has been free from any epileptic attack during the period of one year immediately preceding the date when the licence is granted; or

(b) (if not so free from attack) has had an epileptic attack whilst asleep more than three years before the date when the licence is granted and has had attacks only whilst asleep between the date of that attack and the date when the licence is granted,

where the Secretary of State is satisfied that the driving of a vehicle by him in accordance with the licence is not likely to be a source of danger to the public.

(3) The disability described in paragraph (1) is prescribed for the purposes of section 94(5)(b) of the Traffic Act in relation to an applicant for, or a person who holds, a Group 1 licence.

Disabilities prescribed in respect of Group 2 licences

73.—(1) There is prescribed for the purposes of section 92(2) of the Traffic Act as a relevant disability in relation to an applicant for, or a person who holds, a Group 2 licence, the inability to read in good light (with the aid of corrective lenses if necessary) a registration mark fixed to a motor vehicle and containing letters and figures 79.4 millimetres high at a distance of 20.5 metres.

(2) There is also prescribed for the purposes of section 92(2) of the Traffic Act as a relevant disability in relation to a person other than an excepted licence holder who is an applicant for or who holds a Group 2 licence, such abnormality of sight in one or both eyes that he cannot meet the relevant standard of visual acuity.

(3) The relevant standard of visual acuity for the purposes of paragraph (2) means—

(a) in the case of a person who—

(i) was the holder of a valid Group 2 licence or obsolete vocational licence upon each relevant date specified in column (1) of Table 1 at the end of this regulation, and

(ii) if he is an applicant for a Group 2 licence, satisfies the Secretary of State that he has had adequate recent driving experience and has not during the period of 10 years immediately before the date of the application been involved in any road accident in which his defective eyesight was a contributory factor,

the standard prescribed in relation to him in column (2) of Table 1;

(b) in the case of a person who—

(i) does not fall within sub-paragraph (a), and

(ii) was or is the holder of a valid Group 2 licence upon the relevant date specified in column (1) of Table 2 at the end of this regulation,

the standard prescribed in relation to him in column (2) of Table 2;

(c) in the case of any other person, a standard of visual acuity (with the aid of corrective lenses if necessary) of at least 6/9 in the better and eye and at least 6/12 in the worse eye and, if corrective lenses are necessary, an uncorrected acuity of at least 3/60 in both eyes.

(4) There is prescribed for the purposes of section 92(2) of the traffic Act in relation to a person—

(a) to whom paragraph (3)(c) applies, and

(b) who is able to meet the relevant standard of visual acuity prescribed in that sub-paragraph only with the aid of corrective lenses,

poor toleration of the correction made by the lenses.

(5) There is prescribed for the purposes of section 92(2) as a relevant disability in relation to a person who is an applicant for or who holds a Group 2 licence, sight in only one eye unless—

(a) he held an obsolete vocational licence on 1st April 1991, the traffic commissioner who granted the last such licence knew of the disability before 1st January 1991, and—

(i) in a case of a person who also held such a licence on 1st January 1983, the visual acuity in his sighted eye is no worse than 6/12, or

(ii) in any other case, the visual acuity in his sighted eye is no worse than 6/9, and

if he is an applicant for a Group 2 licence, he satisfies the Secretary of State that he has had adequate recent driving experience and has not during the period of 10 years immediately before the date of the application been involved in any road accident in which his defective eyesight was a contributory factor; or

(b) the person is an excepted licence holder.

(6) Diabetes requiring insulin treatment is prescribed for the purposes of section 92(2) in relation to an applicant for or a person who holds a Group 2 licence unless the person suffering from the disability held an obsolete vocational licence on 1st April 1991 and the traffic commissioner who granted the last obsolete vocational licence knew of the disability before 1st January 1991.

(7) Liability to seizures arising from a cause other than epilepsy is prescribed for the purposes of section 92(2) in relation to an applicant for or a person who holds a Group 2 licence.

(8) Epilepsy is prescribed for the purposes of section 92(4)(b) of the Traffic Act in the case of an applicant for a Group 2 licence suffering from epilepsy who satisfies the Secretary of State that—

(a) during the period of 10 years immediately preceding the date when the licence is granted—

(i) he has been free from any epileptic attack, and

(ii) he has not required any medication to treat epilepsy, and

(b) that the driving of a vehicle by him in accordance with the licence is not likely to be a source of danger to the public.

(9) Diabetes requiring insulin treatment is prescribed for the purposes of section 92(4)(b) in the case of a person who—

(a) is an applicant for a licence authorising the driving of vehicles in sub-category C1 or C1 + E (8.25 tonnes),

(b) held such a licence on 31st December 1996, and

(c) satisfies the Secretary of State that since commencing treatment with insulin, and whilst in employment for the purpose, he has had sufficient recent experience in the driving of vehicles in sub-category C1 or C1 + E (8.25 tonnes) to make practicable an assessment of the risk posed by his driving vehicles of those classes,

provided that he satisfies the conditions mentioned in paragraph (10).

(10) The conditions referred to in paragraph (9) are that—

(a) the Secretary of State was aware on 31 December 1997 that he suffered from that disability,

(b) he has not, during the period of 12 months ending on the date of the application, required the assistance of another person to treat an episode of hypoglycaemia suffered whilst he was driving,

(c) he makes an arrangement to undergo at intervals of not more than 12 months an examination by a hospital consultant specialising in the treatment of diabetes and so far as is reasonably practicable conforms to that arrangement,

(d) his application is supported by a report from such a consultant sufficient to satisfy the Secretary of State that he has a history of responsible diabetic control with a minimal risk of incapacity due to hypoglycaemia during his normal working hours,

(e) he regularly monitors his condition while employed in driving vehicles in sub-category C1 or C1 + E (8.25 tonnes), and

(f) the Secretary of State is satisfied that the driving of such a vehicle in pursuance of the licence is not likely to be a source of danger to the public.

(11) In this regulation—

(a) references to measurements of visual acuity are references to visual acuity measured on the Snellen Scale;

(b) "excepted licence holder" means a person who—

(i) was the holder of a licence authorising the driving of vehicles included in sub-categories C1 and C1 + E (8.25 tonnes) which was in force at a time before 1st January 1997, and

(ii) is an applicant for, or the holder of, a Group 2 licence solely by reason that the licence applied for or held authorises (or would, if granted, authorise) the driving of vehicles included in those sub-categories.

(c) "obsolete vocational licence" means a licence to drive heavy goods vehicles granted under Part IV of the Traffic Act as originally enacted or a licence to drive public service vehicles granted under section 22 of the 1981 Act which was in force a time before 1 April 1991.

TABLE 1

(1) Person holding Group 2 licence or obsolete vocational licence on:	(2) Standard of visual acuity applicable:
1. 1 January 1983 and 1 April 1991	Acuity (with the aid of corrective lenses if necessary) of at least 6/12 in the better eye or at least 6/36 in the worse eye or uncorrected acuity of at least 3/60 in at least one eye.
2. 1 March 1992, but not on 1 January 1983	Acuity (with the aid of corrective lenses if necessary) of at least 6/9 in the better eye or at least 6/12 in the worse eye, or uncorrected acuity of at least 3/60 in at least one eye.

TABLE 2

(1) Person holding Group 2 licence on:	(2) Standard of visual acuity applicable:
1. 31 December 1996, but not on 1 March 1992.	Acuity (with the aid of corrective lenses if necessary) of at least 6/9 in the better eye and at least 6/12 in the worse eye and, if corrective lenses are needed to meet that standard, uncorrected acuity of at least 3/60 in at least one eye.
2. On or after 1 January 1997 but not on 31 December 1996.	Acuity (with the aid of corrective lenses if necessary) of at least 6/9 in the better eye and at least 6/12 in the worse eye and, if corrective lenses are needed to meet that standard, uncorrected acuity of at least 3/60 in both eyes.

Disabilities requiring medical investigation: High Risk Offenders

74.—(1) Subject to paragraph (2), the circumstances prescribed for the purposes of subsection (5) of section 94 of the Traffic Act, under subsection (4) of that section, are that the person who is an applicant for, or holder of, a licence—

    (a) has been disqualified by an order of a court by reason that the proportion of alcohol in his body equalled or exceeded—

        (i) 87.5 microgrammes per 100 millilitres of breath, or

        (ii) 200 milligrammes per 100 millilitres of blood, or

        (iii) 267.5 milligrammes per 100 millilitres of urine;

    (b) has been disqualified by order of a court by reason that he has failed, without reasonable excuse, to provide a specimen when required to do so pursuant to section 7 of the Traffic Act; or

    (c) has been disqualified by order of a court on two or more occasions within any period of 10 years by reason that—

        (i) the proportion of alcohol in his breath, blood or urine exceeded the limit prescribed by virtue of section 5 of the Traffic Act, or

        (ii) he was unfit to drive through drink contrary to section 4 of that Act.

(2) For the purposes of paragraph (1)(a) and (b) a court order shall not be taken into account unless it was made on or after 1st June 1990 and paragraph (1)(c) shall not apply to a person unless the last such order was made on or after 1st June 1990.

Examination by an officer of the Secretary of State

75.—(1) There are prescribed for the purposes of section 94(5)(b)(ii) (examination of a licence applicant or holder by an officer of the Secretary of State) the following disabilities—

    (a) impairment of visual acuity or of the central or peripheral visual field;

    (b) a disability consisting of any one or more of the following—

        (i) the absence of one or more limbs,

        (ii) the deformity of one or more limbs,

        (iii) the loss of use of one or more limbs whether or not progressive in nature, and

        (iv) impairment of co-ordination of movement of the limbs or of co-ordination between a limb and the eye;

    (c) impairment of cognitive functions or behaviour;

(2) In paragraph (1)(b), a reference to a limb includes a reference to part of a limb, and the reference to loss of use in relation to a limb includes a reference to impairment of limb movement, power or sensation.

PART VII

SUPPLEMENTARY

Transitional provisions

Effect of change in classification of vehicles for licensing purposes

76.—(1) In a licence (whether full or provisional) granted before 1st January 1997, a reference to motor vehicles in an old category shall be construed as a reference to motor vehicles in the new category corresponding thereto and a reference to motor vehicles of a class included in an old category shall be construed as a reference to vehicles of the corresponding class included in the new category.

(2) Where a licence granted before 1st January 1997 authorises only the driving of a class of motor vehicles included in an old category having automatic transmission, it shall authorise the driving of the corresponding class of vehicles in the new category having automatic transmission.

(3) For the purposes of paragraphs (1) and (2), a reference in a licence to motor vehicles in an old category (or a class included in that category) includes a reference in a licence granted before 1st June 1990 to a group or class of motor vehicles which is, by virtue of any enactment, to be construed as a reference to vehicles in the old category (or a class included in that category).

(4) In this regulation—

"old category" and "class included in an old category" mean respectively a category and a class of vehicles specified in column (1) of the table at the end of this regulation,

"new category" and "class included in a new category", in relation to an old category, mean respectively the category (or, as the case may be, the sub-category) and the class of vehicles specified in column (2) of the table as corresponding to the relevant old category or class included therein, and

"section 19 permit" means a permit granted under section 19 of the 1985 Act.

TABLE

(1) Old category or class	*(2) Corresponding new category or class*
A	A
B1	B1
B1, limited to invalid carriages	B1 (invalid carriages)
B	B
B plus E	B + E
C1	C1
C1 plus E	C1 + E (8.25 tonnes)
C	C
C plus E	C + E
C plus E, limited to drawbar trailer combinations only	Vehicles in category C + E which are drawbar trailer combinations
D1	D1 (not for hire or reward)
D1 plus E	D1 + E (not for hire or reward)
D, limited to 16 seats	D1
D, limited to vehicles not more than 5.5 metres in length	D1 and vehicles in category D not more than 5.5 metres in length
D, limited to vehicles not driven for hire or reward	Vehicles in category D which are either driven while being used in accordance with a section 19 permit or, if not being so used, driven otherwise than for hire or reward
D	D
D plus E	D + E
F	F
G	G
H	H
K	K
L	L
P	P

Saving in respect of entitlement to Group M

77.—(1) Where a person was authorised by virtue of regulations revoked by these Regulations (whether or not he is also the holder of a licence granted before 1st October 1982) to drive, or to apply for the grant of a licence authorising the driving of, vehicles of a class included in the former group M (trolley vehicles used for the carriage of passengers with more than 16 seats in addition to the driver's seat), he shall continue to be so authorised and any licence granted to such a person shall be construed as authorising the driving of vehicles of that class.

(2) A person who is authorised to drive vehicles of a class included in the former group M shall, to the extent that he is so authorised, be deemed to be the holder of a Group 1 licence.

Saving in respect of entitlement to former category N

78.—(1) Where on 31st December 1996 a person was, by virtue of regulations then in force, the holder of, or entitled to apply for the grant of, a licence authorising the driving of vehicles included in—

(a) the former category N (vehicles exempt from vehicle excise duty under section 7(1) of the Vehicles (Excise) Act 1971) alone, or

(b) category F or A and the former category N,

the Secretary of State may, notwithstanding anything otherwise contained in these Regulations, grant to such a person a licence authorising the driving of vehicles in the former category N (with or without vehicles in either or both of the other categories as the case may be) and a person holding such a licence shall be authorised to drive such vehicles.

(2) Where on 31st December 1996 a person was the holder of, or entitled to apply for the grant of, a licence authorising the driving of vehicles included in category B and the former category N, he shall continue to be authorised to drive vehicles in that former category and any licence granted to such a person authorising the driving of vehicles included in category B shall be construed as authorising also the driving of vehicles in that former category.

Saving in respect of entitlement to drive mobile project vehicles

79. In relation to a person who was at a time before 1st January 1997 the holder of a licence authorising the driving of vehicles of a class included in category B (except a licence authorising only the driving of vehicles included in sub-category B1 or B1 (invalid carriages)), regulation 7(5) shall apply as if paragraphs (b) and (c) and the words "on behalf of a non-commercial body" were omitted.

Miscellaneous

Persons who become resident in Great Britain

80.—(1) A person who becomes resident in Great Britain who is—

(a) the holder of a relevant permit, and

(b) not disqualified for holding or obtaining a licence in Great Britain

shall, during the period of one year after he becomes so resident, be treated for the purposes of section 87 of the Traffic Act as the holder of a licence authorising him to drive all classes of small vehicle, motor bicycle or moped which he is authorised to drive by that permit.

(2) A person who becomes resident in Great Britain who is—

(a) the holder of a British external licence granted in the Isle of Man or Jersey authorising the driving of large goods vehicles of any class, and

(b) not disqualified for holding or obtaining a licence in Great Britain

shall, during the period of one year after he becomes so resident, be treated for the purposes of section 87 of the Traffic Act as the holder of a licence authorising him to drive large goods vehicles of all classes which he is authorised to drive by that licence.

(3) A person who becomes resident in Great Britain who is—

    (a) the holder of a British external licence granted in the Isle of Man or Jersey authorising the driving of passenger-carrying vehicles of any class, and

    (b) not disqualified for holding or obtaining a licence in Great Britain

shall, during the period of one year after he becomes so resident, be treated for the purposes of section 87 of the Traffic Act as the holder of a licence authorising him to drive passenger-carrying vehicles of all classes which he is authorised to drive by that licence.

(4) The enactments mentioned in paragraph (5) shall apply in relation to—

    (a) holders of relevant permits and holders of British external licences of the classes mentioned in paragraphs (2) and (3), or

    (b) (as the case may be) those licences and permits,

with the modifications contained in paragraph (5).

(5) The modifications referred to in paragraph (4) are that—

    (a) section 7 of the Offenders Act(**a**) shall apply as if—

        (i) the references to a licence were references to a relevant permit or a British external licence, and

        (ii) the words after paragraph (c) thereof were omitted;

    (b) section 27(1) and (3) of the Offenders Act(**b**) shall apply as if—

        (i) the references to a licence were references to a relevant permit or a British external licence,

        (ii) the references to the counterpart of a licence were omitted, and

        (iii) in subsection (3) the words ", unless he satisfies the Court that he has applied for a new licence and has not received it" were omitted;

    (c) section 42(5) of the Offenders Act shall apply as if for the words "endorsed on the counterpart of the licence" onwards there were substituted the words "notified to the Secretary of State";

    (d) section 47 of the Offenders Act shall apply as if for subsection (2)(**c**) there were substituted—

    "(2) Where a court orders the holder of a relevant permit or a British external licence to be disqualified it must send the permit or the licence, on its being produced to the court, to the Secretary of State who shall keep it until the disqualification has expired or been removed or the person entitled to it leaves Great Britain and in any case has made a demand in writing for its return to him.

    "Relevant permit" has the meaning given by regulation 80 of the Motor Vehicles (Driving Licences) Regulations 1999.";

    (e) section 164(1), (6) and (8) of the Traffic Act (**d**) shall apply as if the references therein to a licence were references to a relevant permit or a British external licence and the references to a counterpart of a licence were omitted; and

    (f) section 173 of the Traffic Act(**e**) shall apply as if after paragraph (aa) there were added—

    "(ab) a relevant permit (within the meaning of regulation 80 of the Motor Vehicles (Driving Licences) Regulations 1999,

    (ac) a British external licence,".

(6) In this regulation "relevant permit" means—

    (i) a "domestic driving permit",

(**a**) The relevant amendment is by the 1991 Act, Schedule 4, paragraph 83.
(**b**) The relevant amendments are by the 1990 Regulations and the 1991 Act, Schedule 4, paragraph 91.
(**c**) Section 47(2) was amended by the 1990 Regulations and the 1991 Act, Schedule 4, paragraph 100.
(**d**) The relevant amendments are by the 1990 Regulations and the 1991 Act, Schedule 4, paragraph 68, and Schedule 8.
(**e**) The relevant amendment is by the 1996 Regulations.

    (ii) a "Convention driving permit", or

    (iii) a "British Forces (BFG) driving licence",

within the meaning of article 2(7)—of the Motor Vehicles (International Circulation) Order 1975(**a**) which is—

    (a) for the time being valid for the purposes for which it was issued, and

    (b) is not a domestic driving permit or a British Forces (BFG) driving licence in respect of which any order made, or having effect as if made, by the Secretary of State is for the time being in force under article 2(6) of that Order.

Service personnel

81. The traffic commissioner for the South Eastern and Metropolitan Traffic Areas is hereby prescribed for the purposes of section 183(6) of the Traffic Act (discharge of Part IV functions in relation to HM Forces).

Northern Ireland licences

82.—(1) The traffic commissioner for the North Western Traffic Area is hereby prescribed for the purposes of section 122(2)(**b**) of the Traffic Act.

(2) For the purposes of section 122(4) of the Traffic Act, the magistrates' court or sheriff to whom an appeal shall lie by the holder of a Northern Ireland licence, being a person who is not resident in Great Britain and who is aggrieved by the suspension or revocation of the licence or by the ordering of disqualification for holding or obtaining a licence, shall be—

    (a) such a magistrates' court or sheriff as he may nominate at the time he makes the appeal; or

    (b) in the absence of a nomination of a particular court under sub-paragraph (a), the magistrates' court in whose area the office of the traffic commissioner for the North Western Traffic Area is situated.

Statement of date of birth

83.—(1) The circumstances in which a person specified in section 164(2) of the Traffic Act shall, on being required by a police constable, state his date of birth are—

    (a) where that person fails to produce forthwith for examination his licence on being required to do so by a police constable under that section; or

    (b) where, on being so required, that person produces a licence—

        (i) which the police constable in question has reason to suspect was not granted to that person, was granted to that person in error or contains an alteration in the particulars entered on the licence (other than as described in paragraph (ii)) made with intent to deceive; or

        (ii) in which the driver number has been altered, removed or defaced;

    (c) where that person is a person specified in subsection (1)(d) of that section and the police constable has reason to suspect that he is under 21 years of age.

(2) In paragraph (1), "driver number" means the number described as the driver number in the licence.

Signed by authority of the Secretary of State for the Environment, Transport and the Regions

Larry Whitty
Parliamentary under-Secretary of State,
18th October 1999   Department of the Environment, Transport and the Regions

We approve the making of these Regulations

Jim Dowd
Bob Ainsworth
19th October 1999   Two of the Lords Commissioners of Her Majesty's Treasury

---

(**a**) S.I. 1975/1208.
(**b**) Section 122 substituted by section 2(1) of the 1991 Act.

SCHEDULE 1

Regulation 2

Regulations Revoked

Title	Year and Number
The Motor Vehicles (Driving Licences) Regulations 1996	1996/2824
The Motor Vehicles (Driving Licences) (Amendment) Regulations 1997	1997/256
The Motor Vehicles (Driving Licences) (Amendment) (No. 2) Regulations 1997	1997/669
The Motor Vehicles (Driving Licences) (Amendment) (No. 3) Regulations 1997	1997/846
The Motor Vehicles (Driving Licences) (Amendment) (No. 4) Regulations 1997	1997/2070
The Motor Vehicles (Driving Licences) (Amendment) (No. 5) Regulations 1997	1997/2915
The Motor Vehicles (Driving Licences) (Amendment) Regulations 1998	1998/20
The Motor Vehicles (Driving Licences) (Amendment) (No. 2) Regulations 1998	1998/528
The Motor Vehicles (Driving Licences) (Amendment) (No. 3) Regulations 1998	1998/1229
The Motor Vehicles (Driving Licences) (Amendment) (No. 4) Regulations 1998	1998/2038
The Motor Vehicles (Driving Licences) (Amendment) Regulations 1999	1999/72
The Motor Vehicles (Driving Licences) (Amendment) (No. 2) Regulations 1999	1999/617

SCHEDULE 2

Regulations 4 to 6 and 43

CATEGORIES AND SUB-CATEGORIES OF VEHICLE FOR LICENSING PURPOSES

PART 1

(1) Category or sub-category	(2) Classes of vehicle included	(3) Additional categories and sub-categories
A	Motor bicycles.	B1, K and P
A1	A sub-category of category A comprising learner motor bicycles.	P
B	Motor vehicles, other than vehicles included in category A, F, K or P, having a maximum authorised mass not exceeding 3.5 tonnes and not more than eight seats in addition to the driver's seat, including: (i) a combination of any such vehicle and a trailer where the trailer has a maximum authorised mass not exceeding 750 kilogrammes, and (ii) a combination of any such vehicle and a trailer where the maximum authorised mass of the combination does not exceed 3.5 tonnes and the maximum authorised mass of the trailer does not exceed the unladen weight of the tractor vehicle.	F, K and P

65

(1) Category or sub-category	(2) Classes of vehicle included	(3) Additional categories and sub-categories
B1	A sub-category of category B comprising motor vehicles having three or four wheels and an unladen weight not exceeding 550 kilograms.	K and P
B+E	Combinations of a motor vehicle and trailer where the tractor vehicle is in category B but the combination does not fall within that category.	None
C	Motor vehicles having a maximum authorised mass exceeding 3.5 tonnes, other than vehicles falling within category D, F, G or H, including any such vehicle drawing a trailer having a maximum authorised mass not exceeding 750 kilograms.	None
C1	A sub-category of category C comprising motor vehicles having a maximum authorised mass exceeding 3.5 tonnes but not exceeding 7.5 tonnes, including any such vehicle drawing a trailer having a maximum authorised mass not exceeding 750 kilograms.	None
D	Motor vehicles constructed or adapted for the carriage of passengers having more than eight seats in addition to the driver's seat, including any such vehicle drawing a trailer having a maximum authorised mass not exceeding 750 kilograms.	None
D1	A sub-category of category D comprising motor vehicles having more than eight but not more than 16 seats in addition to the driver's seat and including any such vehicle drawing a trailer with a maximum authorised mass not exceeding 750 kilograms.	None
C+E	Combinations of a motor vehicle and trailer where the tractor vehicle is in category C but the combination does not fall within that category.	B+E
C1+E	A sub-category of category C+E comprising combinations of a motor vehicle and trailer where: (a) the tractor vehicle is in sub-category C1, (b) the maximum authorised mass of the trailer exceeds 750 kilograms but not the unladen weight of the tractor vehicle, and (c) the maximum authorised mass of the combination does not exceed 12 tonnes.	B+E
D+E	Combinations of a motor vehicle and trailer where the tractor vehicle is in category D but the combination does not fall within that category.	B+E
D1+E	A sub-category of category D+E comprising combinations of a motor vehicle and trailer where: (a) the tractor vehicle is in sub-category D1, (b) the maximum authorised mass of the trailer exceeds 750 kilograms but not the unladen weight of the tractor vehicle, (c) the maximum authorised mass of the combination does not exceed 12 tonnes, and (d) the trailer is not used for the carriage of passengers.	B+E
F	Agricultural or forestry tractors, including any such vehicle drawing a trailer but excluding any motor vehicle included in category H.	K
G	Road rollers.	None
H	Track-laying vehicles steered by their tracks.	None
K	Mowing machines which do not fall within category A and vehicles controlled by a pedestrian.	None
P	Mopeds.	None

(1) Sub-category	(2) Classes of vehicle included	(3) Additional categories and sub-categories
C1 + E (8.25 tonnes)	A sub-category of category C + E comprising combinations of a motor vehicle and trailer in sub-category C1 + E, the maximum authorised mass of which does not exceed 8.25 tonnes.	None
D1 (not for hire or reward)	A sub-category of category D comprising motor vehicles in sub-category D1 driven otherwise than for hire or reward.	None
D1 + E (not for hire or reward)	A sub-category of category D + E comprising motor vehicles in sub-category D1 + E driven otherwise than for hire or reward.	None
L	Motor vehicles propelled by electrical power.	None

PART 3

(1) Sub-category	(2) Classes of vehicle included	(3) Additional categories and sub-categories
B1 (invalid carriages)	A sub-category of category B comprising motor vehicles which are invalid carriages.	None

SCHEDULE 3   Regulation 14

LICENCE FEES

PART 1

TABLE OF FEES

No.	Description of Licence and circumstances of application	Fee payable
1.	A first licence.	£23.50
2.	A provisional licence granted following the revocation of a licence under the Road Traffic (New Drivers) Act 1995.	£23.50
3.	A full licence granted in exchange for a full Northern Ireland licence or to a person who has held a full Northern Ireland licence which was granted on or after 1st January 1976.	£13.50
4.	A provisional licence authorising the driving of vehicles of a class included in category A except— (a) a first licence (b) a licence granted following the revocation of a licence under section 93(1) or (2) of the Traffic Act or the delivery of a Community licence to the Secretary of State under section 99C(1), (2) or (3) of the Traffic Act(**a**), or (c) a licence falling within paragraph 8, 12 or 13.	£13.50

(**a**) Section 99C was inserted by the 1996 Regulations.

No.	Description of Licence and circumstances of application	Fee payable
5.	A licence which is the applicant's first full Group 1 licence, first full licence authorising the driving of vehicles in category C or first full licence authorising the driving of vehicles in category D.	£8.50
6.	A licence granted in exchange for a licence still in force or in place of a licence which has been revoked or a Community licence which is required to be delivered to the Secretary of State, except—	
	(a) a licence granted pursuant to section 118(4) of the Traffic Act,	
	(b) a licence granted following the revocation of a licence under section 93(1) of the Traffic Act or the delivery of a Community licence to the Secretary of State under section 99C(1) or (3) of the Traffic Act,	
	(c) a licence which the Secretary of State is required to grant free of charge under section 93(2), 99(7) or 99C(2) of the Traffic Act,	
	(d) a short Group 2 licence, or	
	(e) a licence falling within any other paragraph of this Part.	£13.50
7.	A Group 1 licence granted upon the expiry of a previous Group 1 licence, except—	
	(a) a licence falling within paragraph 4, or	
	(b) a licence granted under section 99(1)(b) of the Traffic Act.	£8.50
8.	A licence granted by way of replacement of a lost or defaced licence.	£13.50
9.	A provisional licence authorising the driving of vehicles of a class included in category C.	£23.50
10.	A provisional licence authorising the driving of vehicles of a class included in category D.	£23.50
11.	A full Group 2 licence granted upon the expiry of a previous full Group 2 licence, other than a licence granted under section 99(1)(b) or (1A)(c) of the Traffic Act(**a**).	£28.50
12.	A licence, other than a short Group 2 licence, granted upon the expiry of a period of disqualification ordered by a court under section 34 or 35 of the Offenders Act—	
	(a) in the circumstances prescribed under section 94(4) of the Traffic Act irrespective of the date when the court order was made;	£33.50
	(b) otherwise than in those circumstances.	£24.50
13.	A provisional licence, other than a first licence, granted following disqualification ordered by a court under section 36 of the Offenders Act(**b**), whether or not the court also made an order under section 34 or 35 of that Act—	
	(a) in the circumstances prescribed under section 94(4) of the Traffic Act(**c**) irrespective of the date when the court order was made;	£33.50
	(b) otherwise than in those circumstances.	£24.50
14.	A provisional Group 2 licence granted first to a person who has been ordered to take a driving test under section 117 or 117A of the Traffic Act(**d**)	£24.50

(**a**) Subsection (1A) was inserted by section 2(2) of the 1989 Act.
(**b**) Section 36 was substituted by section 32 of the Road Traffic Act 1991 (c. 40).
(**c**) The relevant amendment is by section 5(1) of the 1989 Act.
(**d**) Section 117 was substituted by section 2(1) of and Schedule 2 to the 1989 Act. Section 117A was inserted by the 1996 Regulations.

INTERPRETATION

In Part 1 of this Schedule—

"first licence" means a licence (other than a licence falling within paragraph 3 of the Table in Part 1) granted to a person—

(a) who has not held a licence before, or

(b) whose last licence was a full licence which expired before 31st December 1978, or

(c) whose last licence was a provisional licence which was granted before 1st October 1982;

"short Group 2 licence" means a Group 2 licence which, when granted upon the expiry of a period of disqualification, must (by virtue of the expiry date of the licence which was revoked upon disqualification) expire not later than three months after the date it is granted.

SCHEDULE 4

Regulation 16

DISTINGUISHING MARKS TO BE DISPLAYED ON A MOTOR VEHICLE BEING DRIVEN UNDER A PROVISIONAL LICENCE

PART 1

Diagram of distinguishing mark to be displayed on a motor vehicle in England, Wales or Scotland.

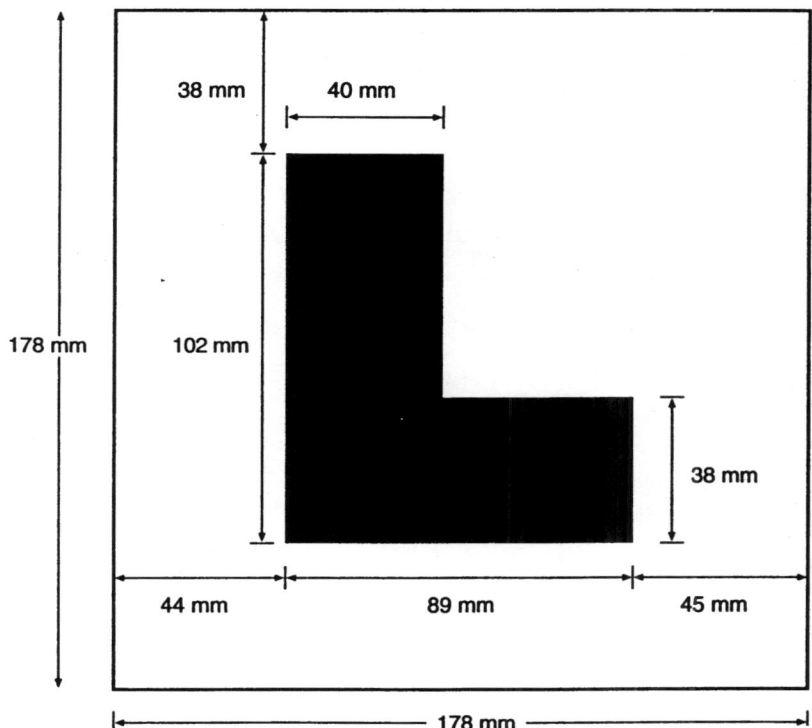

Red letter on white ground.

The corners of the ground can be rounded off.

PART 2

Diagram of optional distinguishing mark to be displayed on a motor vehicle in Wales if a mark in the form set out in Part 1 is not displayed.

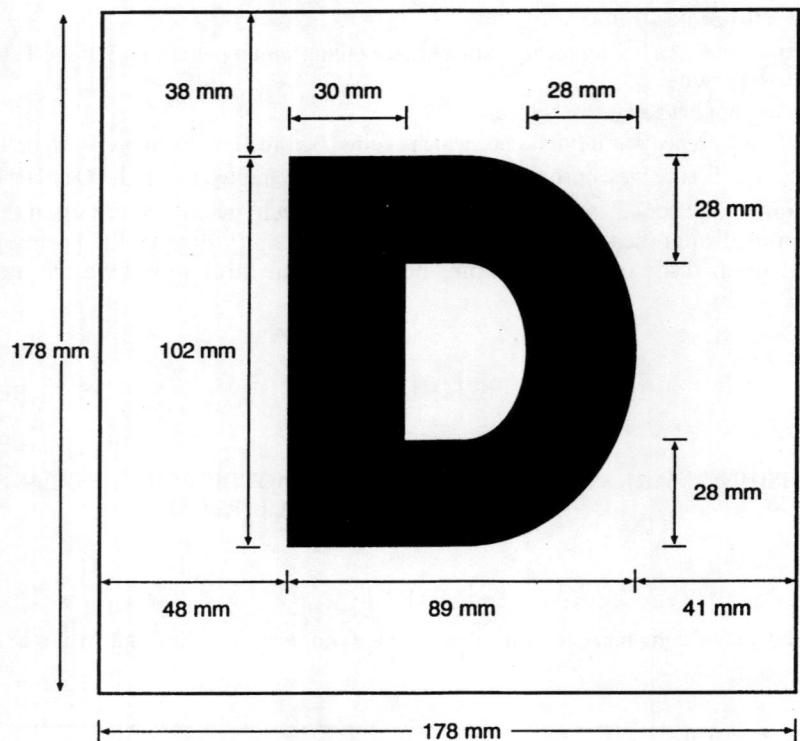

Red letter on white ground.

The corners of the ground can be rounded off.

SCHEDULE 5

Regulation 35

FEES FOR PRACTICAL AND UNITARY TESTS

(1) Category or sub-category of vehicle	(2) Test, other than extended driving test, commencing:		(3) Extended driving test, commencing:	
	(a) During normal hours	(b) Out of hours	(a) During normal hours	(b) Out of hours
1. A and P	£45.00	£55.00	£90.00	£110.00
2. B1, B, F, G, H and K,	£36.75	£46.00	£73.50	£92.00
3. B+E, C1, C1+E, D1, D1+E, C, C+E, D and D+E	£73.50	£92.00	—	—

70

EVIDENCE OF IDENTITY OF TEST CANDIDATES

1. The documents referred to in regulation 38(3) and (5) are:—

    (a) a passport;

    (b) any of the following documents if it bears a photograph and signature of the person, namely—

        (i) a cheque guarantee card or credit card,

        (ii) an employer's identity card,

        (iii) a trade union card,

        (iv) a students union card,

        (v) a school bus pass, or

        (vi) a card issued in connection with the sale and purchase of railway tickets;

    (c) a photograph of the person which has been endorsed with a certificate in the prescribed form signed by an acceptable person.

2. In this Schedule—

"acceptable person" means a bank official, a certified instructor (within the meaning of regulation 60(5)(a)), a commissioned officer in Her Majesty's Forces, an established civil servant, a person whose name is entered in the register of driving instructors under Part V of the Traffic Act, a local authority councillor, a Justice of the Peace, a medical practitioner, a Member of Parliament, a minister of religion, a police officer, a solicitor or barrister or a teacher;

"the prescribed form", in relation to a certificate, means the following—

        "I, [Name of acceptable person], certify that this is a true likeness of [Name of candidate], who has been known to me for [...] years in my capacity as [Specify capacity]

        Signed...

        Date...

    Business or profession (and registration or certificate number, if any) ....

        Telephone number..."

or a form substantially to the same effect.

SPECIFIED MATTERS FOR THEORY TEST

PART I

CATEGORIES A AND P

The specified matters are set out in sections A to G. The person conducting the test shall examine candidates on all the items included in sections A to G but need not examine them on every item mentioned in sections F and G provided that he asks at least one question about them at random.

A. Road traffic regulation

Road traffic regulations including road signs, road markings, signals, rights of way and speed limits.

B. The driver

1. The importance of alertness and attitudes to other road users.

2. Perception, judgement and decision-making, including especially reaction time and changes in driver behaviour due to the influence of alcohol, drugs and medicinal products, state of mind and fatigue.

C. The road

1. The most important principles concerning the observance of safe distance between vehicles, braking distances and roadholding under various weather and road conditions.

2. Driving risk factors related to various road conditions, in particular as they change with the weather and the time of day or night.

3. Characteristics of various types of road and the related statutory requirements.

D. Other road users

1. Specific risk factors related to the lack of experience of other road users and the most vulnerable categories of user such as children, pedestrians, cyclists and people whose mobility is reduced.

2. Risks involved in the movement and driving of various types of vehicle and of the different fields of view of their drivers.

E. General rules and regulations and other matters

1. Rules concerning the administrative documents required for the use of vehicles.

2. General rules specifying how the driver must behave in the event of an accident (setting warning device and raising the alarm) and the measures which he can take to assist road accident victims where necessary.

3. Safety factors relating to persons carried, including balancing with a passenger.

F. Road and Vehicle safety

1. Mechanical aspects of the vehicle with a bearing on road safety, i.e. the detection of the most common faults, in particular in the steering, suspension and brake systems, tyres, lights and direction indicators, reflectors, rear-view mirrors, and the exhaust system.

2. Vehicle safety equipment including, in particular, the use of crash helmets and visors.

G. Environmental matters

Rules regarding vehicle use in relation to the environment, including the appropriate use of audible warning devices, moderate fuel consumption, limitation of pollutant emissions and matters of a similar nature.

PART 2

CATEGORY B

The specified matters are set out in sections A to G. The person conducting the test shall examine candidates on all the items included in sections A to G but need not examine them on every item mentioned in sections F and G provided that he asks at least one question about them at random.

A. Road traffic regultion

Road traffic regulations including road signs, road markings, signals, rights of way and speed limits.

B. The driver

1. The importance of alertness and attitudes to other road users.

2. Perception, judgement and decision-making, including especially reaction time and changes in driver behaviour due to the influence of alcohol, drugs and medicinal products, state of mind and fatigue.

C. The road

1. The most important principles concerning the observance of safe distance between vehicles, braking distances and roadholding under various weather and road conditions.

2. Driving risk factors related to various road conditions, in particular as they change with the weather and the time of day or night.

3. Characteristics of various types of road and the related statutory requirements.

D. Other road users

1. Specific risk factors related to the lack of experience of other road users and the most vulnerable categories of user such as children, pedestrians, cyclists and people whose mobility is reduced.

2. Risks involved in the movement and driving of various types of vehicle and of the different fields of view of their drivers.

E. General rules and regulations and other matters

1. Rules concerning the administrative documents required for the use of vehicles.

2. General rules specifying how the driver must behave in the event of an accident (setting warning device and raising the alarm) and the measures which he can take to assist road accident victims where necessary.

3. Safety factors relating to persons carried.

F. Road and Vehicle safety

1. Mechanical aspects of the vehicle with a bearing on road safety, i.e. the detection of the most common faults, in particular in the steering, suspension and brake systems, tyres, lights and direction indicators, reflectors, rear-view mirrors, windscreen and wipers, and the exhaust system and seat-belts.

2. Vehicle safety equipment including, in particular, the use of seat-belts and child safety equipment.

G. Environmental matters

Rules regarding vehicle use in relation to the environment, including the appropriate use of audible warning devices, moderate fuel consumption, limitation of pollutant emissions and matters of a similar nature.

PART 3

CATEGORY C

The specified matters are set out in sections A to G. The person conducting the test shall examine the candidate on all the items included in those sections.

A. Road traffic regulation

Road traffic regulations including road signs, road markings, signals, rights of way and speed limits.

B. The driver

1. The importance of alertness and attitudes to other road users.

2. Perception, judgement and decision-making, including especially reaction time and changes in driver behaviour due to the influence of alcohol, drugs and medicinal products, state of mind and fatigue.

C. The road

1. The most important principles concerning the observance of safe distance between vehicles, braking distances and roadholding under various weather and road conditions.

2. Driving risk factors related to various road conditions as they change with the weather and the time of day or night, in particular, the effect of wind on the course of the vehicle.

3. Characteristics of various types of road and the related statutory requirements.

4. Precautions to be taken when overtaking because of the danger of splashing spray or mud.

D. Other road users

1. Specific risk factors related to the lack of experience of other road users and the most vulnerable categories of user such as children, pedestrians, cyclists and people whose mobility is reduced.

2. Risks involved in the movement and driving of various types of vehicle.

3. Obstruction of the field of view of the driver and other road users caused by characteristics of their vehicles.

E. General rules and regulations and other matters

1. Rules concerning the administrative documents required for the use of vehicles.

2. General rules specifying how the driver must behave in the event of an accident (setting warning device and raising the alarm) and the measures which he can take to assist road accident victims where necessary.

3. Rules on vehicle weights and dimensions.

4. Rules on driving hours, rest periods and the use of the tachograph.

F. Road and vehicle safety

1. Mechanical aspects of the vehicle with a bearing on road safety, ie checks to detect the most common faults, in particular in the steering, suspension and brake systems, tyres, lights and direction indicators, reflectors, rear-view mirrors, audible warning devices, windscreen and wipers, the exhaust system and seat-belts.

2. Vehicle safety equipment.

3. Principles of braking systems and speed governors.

4. Precautions when alighting from the vehicle.

5. Safety factors relating to vehicle loading.

G. Environmental matters

Rules regarding vehicle use in relation to the environment, including the appropriate use of audible warning devices, moderate fuel consumption, limitation of pollutant emissions and matters of a similar nature.

PART 4

CATEGORY D

The specified matters are set out in sections A to G. The person conducting the test shall examine the candidate on all the items included in those sections.

A. Road traffic regulation

Road traffic regulations including road signs, road markings, signals, rights of way and speed limits.

B. The driver

1. The importance of alertness and attitudes to other road users.

2. Perception, judgement and decision-making, including especially reaction time and changes in driver behaviour due to the influence of alcohol, drugs and medicinal products, state of mind and fatigue.

C. The road

1. The most important principles concerning the observance of safe distance between vehicles, braking distances and roadholding under various weather and road conditions.

2. Driving risk factors related to various road conditions as they change with the weather and the time of day or night, in particular, the effect of wind on the course of the vehicle.

3. Characteristics of various types of road and the related statutory requirements.

4. Precautions to be taken when overtaking because of the danger of splashing spray or mud.

D. Other road users

1. Specific risk factors related to the lack of experience of other road users and the most vulnerable categories of user such as children, pedestrians, cyclists and people whose mobility is reduced.

2. Risks involved in the movement and driving of various types of vehicle.

3. Obstruction of the field of view of the driver and other road users caused by characteristics of their vehicles.

E. General rules and regulations and other matters

1. Rules concerning the administrative documents required for the use of vehicles.

2. General rules specifying how the driver must behave in the event of an accident (setting warning device and raising the alarm) and the measures which he can take to assist road accident victims where necessary.

3. Rules or vehicle weights and dimensions.

4. Rules on driving hours, rest periods and the use of the tachograph.

5. Rules concerning persons carried.

F. Road and vehicle safety

1. Mechanical aspects of the vehicle with a bearing on road safety, ie checks to detect the most common faults, in particular in the steering, suspension and brake systems, tyres, lights and direction indicators, reflectors, rear-view mirrors, audible warning devices, windscreen and wipers, the exhaust system and seat-belts.

2. Vehicle safety equipment.

3. Principles of braking systems and speed governors.

4. Precautions when alighting from the vehicle.

5. Safety factors relating to vehicle loading and persons carried.

G. Environmental matters

Rules regarding vehicle use in relation to the environment, including the appropriate use of audible warning devices, moderate fuel consumption, limitation of pollutant emissions and matters of a similar nature.

SCHEDULE 8   Regulation 40

SPECIFIED REQUIREMENTS FOR PRACTICAL OR UNITARY TEST

PART I

PRACTICAL TEST: CATEGORIES A AND P

Each test candidate must satisfy the person conducting the test as to—
   (a) his ability to—
      (i) carry out properly the activities, and
      (ii) perform competently, without danger to and with due consideration for other road users, the manoeuvres

      specified in sections A to E below in all respects in accordance with those sections; and
   (b) his understanding of how to balance safely with a passenger.

A. Eyesight

Read in good daylight (with the aid of corrective lenses if worn) a registration mark fixed to a motor vehicle and containing letters and figures 79.4 millimetres high at a distance of 20.5 metres.

B. Preparation to drive

1. Adjust rear view mirrors.

2. Adjust crash helmet.

C. Technical control of the vehicle

1. Start the engine and move off smoothly (uphill and downhill as well as on the flat).

2. Accelerate to a suitable speed while maintaining a straight course, including during gear-changes.

3. Adjust speed to negotiate left or right turns at junctions, possibly in restricted spaces, while maintaining control of the vehicle.

4. Lean over to turn.

5. Keep balance at various speeds.

6. Brake accurately to stop where directed, if need be by performing an emergency stop.

7. Park the vehicle on its stand.

8. Remove motor bicycle from its stand and move it, without the aid of the engine, by walking alongside it.

9. Cause the vehicle to face in the opposite direction by driving it forward (a"U-turn").

D. Behaviour in traffic

1. Observe (including the use of the rear view mirrors) road alignment, markings, signs and potential or actual risks.

2. Communicate with other road users using the authorised means.

3. React appropriately in actual risk situations.

4. Comply with road traffic regulations and the instructions of the police and traffic controllers.

5. Move off from the kerb or a parking space.

6. Drive with the vehicle correctly positioned on the road, adjusting speed to traffic conditions and the line of the road.

7. Keep the right distance between vehicles.

8. Change lanes.

9. Pass parked or stationary vehicles and obstacles.

10. Approach and cross junctions.

11. Turn right and left at junctions or to leave the carriageway.

12. Where the opportunity arises—
    (a) Pass oncoming vehicles, including in confined spaces.
    (b) Overtake in various situations.
    (c) Approach and cross level-crossings.

E. Alighting from vehicle

Take all precautions necessary when alighting.

PART 2

PRACTICAL TEST: CATEGORIES B AND B + E

(1) Each test candidate must, subject to paragraphs (2) and (3), satisfy the person conducting the test as to his ability to—
    (i) carry out properly the activities, and
    (ii) perform competently, without danger to and with due consideration for other road users, the manoeuvres

specified in sections A to E below in all respects in accordance with those sections.

(2) Any requirement contained in those paragraphs shall, in the case of a test for a licence authorising the driving of a motor vehicle in sub-category B1, only be complied with to the extent that it is compatible with the characteristics of the vehicle on which the test is taken.

(3) A test candidate undertaking a test for a licence authorising the driving of a vehicle of a class included in category B + E must in addition satisfy the person conducting the test as to his ability to carry out properly the activity specified in section F but in the case of a disabled driver, this requirement may be carried out through oral questioning.

A. Eyesight

Read in good daylight (with the aid of corrective lenses if worn) a registration mark fixed to a motor vehicle and containing letters and figures 79.4 millimetres high at a distance of 20.5 metres.

B. Preparation to drive

1. Adjust the seat as necessary to obtain a correct seating position.

2. Adjust rear view mirrors and seat belt.

3. Check that the doors are closed.

C. Technical control of the vehicle

1. Start the engine and move off smoothly (uphill and downhill as well as on the flat).

2. Accelerate to a suitable speed while maintaining a straight course, including during gear-changes.

3. Adjust speed to negotiate left or right turns at junctions, possibly in restricted spaces, while maintaining control of the vehicle.

4. Brake accurately to stop where directed, if need be by performing an emergency stop.

5. Either—
    (a) perform any two of the following manoeuvres—
        (i) reverse in a straight line and reverse right or left round a corner while keeping within the correct traffic lane;
        (ii) turn the vehicle to face the opposite way, using forward and reverse gears;
        (iii) park the vehicle and leave a parking space (parallel, oblique or right-angle) both forwards and in reverse, on the flat, uphill and downhill; or
    (b) (in the case of a test for a licence authorising the driving of vehicles in category B + E only) reverse in an S-shaped curve.

D. Behaviour in traffic

1. Observe (including the use of the rear view mirrors) road alignment, markings, signs and potential or actual risks.

2. Communicate with other road users using the authorised means.

3. React appropriately in actual risk situations.

4. Comply with road traffic regulations and the instructions of the police and traffic controllers.

5. Move off from the kerb or a parking space.

6. Drive with the vehicle correctly positioned on the road, adjusting speed to traffic conditions and the line of the road.

7. Keep the right distance between vehicles.

8. Change lanes.

9. Pass parked or stationary vehicles and obstacles.

10. Approach and cross junctions.

11. Turn right and left at junctions or to leave the carriageway.

12. Where the opportunity arises—
    (a) Pass oncoming vehicles, including in confined spaces.
    (b) Overtake in various situations.
    (c) Approach and cross level-crossings.

E. Alighting from vehicle
Take all precautions necessary when alighting.

F. Trailer
Uncouple and recouple trailer.

PART 3

PRACTICAL TEST: CATEGORIES C AND C + E

(1) Each test candidate must satisfy the person conducting the test as to his ability to—
    (i) carry out properly the activities, and
    (ii) perform competently, without danger to and with due consideration for other road users, the manoeuvres

prescribed in paragraph (2), (3) or (4), as the case may be, in all respects in accordance with this Part of this Schedule.

(2) In the case of candidates taking a test for a licence authorising the driving of vehicles of a class included in category C or C + E or any sub-category thereof, the manoeuvres and activities mentioned in sections A to C and F below are prescribed.

(3) In the case of candidates taking a test for a licence authorising the driving of vehicles of a class included in category C or C + E or in a sub-category C1 + E, the manoeuvres and activities mentioned in section D are also prescribed.

(4) In the case of candidates taking a test for a licence authorising the driving of vehicles of a class included in category C + E or sub-category C1 + E, the activity mentioned in section E is also prescribed.

A. Preparation to drive

1. Adjust the seat as necessary to obtain a correct seating position.

2. Adjust rear view mirrors and seat belt.

3. Check that the doors are closed.

B. Technical control of the vehicle

1. Start the engine and move off smoothly (uphill and downhill as well as on the flat).

2. Accelerate to a suitable speed while maintaining a straight course, including during gear-changes.

3. Adjust speed to negotiate left or right turns at junctions, possibly in restricted spaces, while maintaining control of the vehicle.

4. Brake accurately to stop where directed, if need be by performing an emergency stop.

5. Either—
    (a) perform any two of the following manoeuvres—
        (i) reverse in a straight line and reverse right or left round a corner while keeping within the correct traffic lane;
        (ii) turn the vehicle to face the opposite way, using forward and reverse gears;
        (iii) park the vehicle and leave a parking space (parallel, oblique or right-angle) both forwards and in reverse, on the flat, uphill and downhill; or
    (b) reverse in an S-shaped curve.

C. Behaviour in traffic

1. Observe (including the use of the rear view mirrors) road alignment, markings, signs and potential or actual risks.

2. Communicate with other road users using the authorised means.

3. React appropriately in actual risk situations.

4. Comply with road traffic regulations and the instructions of the police and traffic controllers.

5. Move off from the kerb or a parking space.

6. Drive with the vehicle correctly positioned on the road, adjusting speed to traffic conditions and the line of the road.

7. Keep the right distance between vehicles.

8. Change lanes.

9. Pass parked or stationary vehicles and obstacles.

10. Approach and cross junctions.

11. Turn right and left at junctions or to leave the carriageway.

12. Where the opportunity arises—
    (a) Pass oncoming vehicles, including in confined spaces.
    (b) Overtake in various situations.
    (c) Approach and cross level-crossings.

D. Larger vehicles: speed reduction and steering

1. Check the power-assisted braking and steering systems.

2. Use the various braking systems.

3. Use the speed reduction systems other than the brakes.

4. Adjust course to allow for the length of the vehicle and its overhang.

E. Trailers

Uncouple and re-couple trailer or semi-trailer from and to the tractor vehicle.

F. Vehicle safety

Show awareness of vehicle safety measures and be able to operate vehicle safety systems.

PART 4

PRACTICAL TEST: CATEGORIES D AND D + E

(1) Each test candidate must satisfy the person conducting the test as to his ability to—

    (i) carry out properly the activities, and

    (ii) perform competently, without danger to and with due consideration for other road users, the manoeuvres

prescribed in paragraph (2), (3) or (4), as the case may be, in all respects in accordance with this Part of this Schedule.

(2) In the case of candidates taking a test for a licence authorising the driving of vehicles of a class included in category D or D + E or any sub-category thereof, the manoeuvres and activities mentioned in sections A to C and F below are prescribed.

(3) In the case of candidates taking a test for a licence authorising the driving of vehicles of a class included in category D or D + E or in sub-category D1 + E, the manoeuvres and activities mentioned in section D are also prescribed.

(4) In the case of candidates taking a test for a licence authorising the driving of vehicles of a class included in category D + E or sub-category D1 + E, the activity mentioned in section E is also prescribed.

A. Preparation to drive

1. Adjust the seat as necessary to obtain a correct seating position.

2. Adjust rear view mirrors and seat belt.

3. Check that the doors are closed.

B. Technical control of the vehicle

1. Start the engine and move off smoothly (uphill and downhill as well as on the flat).

2. Accelerate to a suitable speed while maintaining a straight course, including during gear-changes.

3. Adjust speed to negotiate left or right turns at junctions, possibly in restricted spaces, while maintaining control of the vehicle.

4. Brake accurately to stop where directed, if need be by performing an emergency stop.

5. Either—

    (a) perform any two of the following manoeuvres—

        (i) reverse in a straight line and reverse right or left around a corner while keeping within the correct traffic lane;

        (ii) turn the vehicle to face the opposite way, using forward and reverse gears;

        (iii) park the vehicle and leave a parking space (parallel, oblique or right-angle) both forwards and in reverse, on the flat, uphill and downhill; or

    (b) reverse in an S-shaped curve.

C. Behaviour in traffic

1. Observe (including the use of the rear view mirrors) road alignment, markings, signs and potential or actual risks.

2. Communicate with other road users using the authorised means.

3. React appropriately in actual risk situations.

4. Comply with road traffic regulations and the instructions of the police and traffic controllers.

5. Move off from the kerb or a parking space.

6. Drive with the vehicle correctly positioned on the road, adjusting speed to traffic conditions and the line of the road.

7. Keep the right distance between vehicles.

8. Change lanes.

9. Pass parked or stationary vehicles and obstacles.

10. Approach and cross junctions.

11. Turn right and left at junctions or to leave the carriageway.

12. Where the opportunity arises—
    (a) Pass oncoming vehicles, including in confined spaces.
    (b) Overtake in various situations.
    (c) Approach and cross level-crossings.

D. Larger vehicles: speed reduction and steering

1. Check the power-assisted braking and steering systems.

2. Use the various braking systems.

3. Use the speed reduction systems other than the brakes.

4. Adjust course to allow for the length of the vehicle and its overhang.

E. Trailers

Uncouple and re-couple trailer or semi-trailer from and to the tractor vehicle.

F. Vehicle safety

Show awareness of vehicle safety measures and be able to operate vehicle safety systems.

PART 5

UNITARY TEST: CATEGORIES F, G, H AND K

1. Read in good daylight (with the aid of corrective lenses if worn) a registration mark fixed to a motor vehicle and containing letters and figures 79.4 millimetres high at a distance of :
    (a) 12.3 metres, in the case of a unitary test conducted in respect of a vehicle included in category K;
    (b) 20.5 metres in any other case.

2. Start the engine of the vehicle.

3. Move away straight ahead or at an angle.

4. Overtake, meet or cross the path of other vehicles and take an appropriate course.

5. Turn right-hand and left-hand corners correctly.

6. Stop the vehicle in an emergency and normally and, in the latter case, bring it to rest in an appropriate part of the road.

PART 6

UNITARY TEST: CATEGORIES F AND G

1. Carry out manoeuvres involving the use of reverse gear (except in a case where the vehicle is not fitted with a means of reversing).

2. Indicate intended actions at appropriate times by giving appropriate signals in a clear and unmistakable manner.

In the case of a test taken on a vehicle with a left-hand drive or by a disabled person for whom it is impracticable or undersirable to give signals by arm, there shall be no requirement to give signals which cannot be given by mechanical means.

3. Act correctly and promptly on all signals given by traffic signs and traffic controllers and take appropriate action in relation to signs given by other road users.

PART 7

UNITARY TEST: CATEGORY H

1. Indicate intended actions at appropriate times by giving appropriate signals in a clear and unmistakable manner.

In the case of a test taken on a vehicle with a left-hand drive or by a disabled person for whom it is impracticable or undersirable to give signals by arm, there shall be no requirement to give signals which cannot be given by mechanical means.

2. Act correctly and promptly on all signals given by traffic signs and traffic controllers and take appropriate action in relation to signs given by other road users.

3. Drive the vehicle backwards and cause it to face in the opposite direction by means of its tracks.

SCHEDULE 9   Regulation 45

UPGRADED ENTITLEMENTS ON PASSING SECOND TEST

TABLE A

(1) Test prescribed in respect of:—	Prescribed test also passed for:—	
	(2) Category C+E	(3) Sub-category C1+E
D	D+E	D1+E
D1	D1+E	D1+E

TABLE B

(A) Automatic test pass:—	Manual test pass in category (or sub-category):—							
	(1) C1	(2) C	(3) C1+E	(4) C+E	(5) D1	(6) D	(7) D1+E	(8) D+E
C1	—	—	C1	C1 & C1+E	C1	C1	C1	C1
C	—	—	—	C	C1	C	C1	C
C1+E	—	C1+E	—	—	D1+E (a)	C1+E	C1+E	C1+E
C+E	—	—	—	—	D1+E (a)	D+E (a)	—	C+E
D1	D1	D1	D1 & D1+E	D1 & D1+E	—	—	—	D1 & D1+E
D	—	D	—	D & D+E	—	—	—	D
D1+E	—	D1+E	D1+E	D1+E	—	D1+E	—	—
D+E	—	—	—	D+E	—	—	—	—

FORMS OF CERTIFICATE AND STATEMENT OF THEORY TEST RESULT

PART 1

THEORY TEST PASS CERTIFICATE

Certificate of passing a theory test

Driver number...

Date of test...

It is hereby certified that [Name of candidate] has been examined and has PASSED the theory test prescribed under section 89 of the Road Traffic Act 1988 in respect of category/categories......................

Theory Test Centre [Number or location of centre].

PART 2

THEORY TEST FAILURE STATEMENT

Statement of failure to pass a theory test

Driver number...

Date of test...

[Name of candidate] has been examined and has FAILED to pass the theory test prescribed under section 89 of the Road Traffic Act 1988 in respect of category/categories...

Theory Test Centre [Number or location of centre].

SCHEDULE 11   Regulation 48

FORMS OF CERTIFICATE AND STATEMENT OF PRACTICAL AND UNITARY TEST RESULT

PART 1

TEST PASS CERTIFICATE

Certificate of passing the [*extended**] test of competence to drive

Driver number...

Date of test...

Test Centre...

I certify that [Name of candidate] has been examined and has PASSED the test of competence prescribed for the purposes of section 89 of the Road Traffic Act [*and section 36 of the Road Traffic Offenders Act**] 1988 in respect of vehicles in category/categories ....

Whether vehicle fitted with automatic transmission ....Y/N.

Whether vehicle modified/other restrictions ....

Signature of examiner.......................................

Signature of candidate.....................................

*Words in italics to be omitted where inapplicable.

PART 2

TEST FAILURE STATEMENT

Statement of failure to pass the practical test/test of competence to drive

(To be endorsed on the front or the reverse of the Driving Test Report Form)

Name of candidate ....

Category/ies of vehicle.....................................

Driver Number...

Date of test...

The candidate named herein has been examined and has FAILED to pass the practical test/test of competence to drive prescribed under the Road Traffic Act [*and for the purposes of section 36 of the Road Traffic Offenders Act*] 1988 in respect of vehicles in the above category/categories.

*Words in italics to be omitted where inapplicable.

ELEMENTS OF AN APPROVED TRAINING COURSE

(A) Introduction

1. Trainees must be told and must understand:—

— the aims of the approved training course;

— the importance of having the right equipment and clothing.

2. Trainees' eyesight must be tested. Trainees must be able to read, in good daylight, a vehicle registration mark containing letters and figures 79.4 mm high at a distance of 20.5 metres (with the aid of glasses or contact lenses if worn).

(B) Practical on site training

Trainees must receive practical on site training at the conclusion of which they must fulfil the following requirements, that is to say they must:—

— be familiar with the motor cycle, its controls and how it works;

— be able to carry out basic machine checks to a satisfactory standard and be able to take the bike on and off the stand satisfactorily;

— be able to wheel the machine around to the left and right showing proper balance and bring the motorcycle to a controlled halt by braking;

— be able to start and stop the engine satisfactorily.

(C) Practical on site riding

Trainees must undertake practical on site riding at the conclusion of which they must be able to:—

— ride the machine under control in a straight line and bring the machine to a controlled halt;

— ride the machine round a figure of eight circuit under control;

— ride the machine slowly under control;

— carry out a U-turn manoeuvre satisfactorily;

— bring the machine to a stop under full control as in an emergency;

— carry out controlled braking using both brakes;

— change gear satisfactorily;

— carry out rear observation correctly;

— carry out simulated left and right hand turns correctly using the Observation-Signal-Manoeuvre (OSM) and Position-Speed-Look (PSL) routines.

(D) Practical on road training

1. Before undertaking practical on road riding trainees must be instructed in the matters set out in paragraphs 2 and 3 below and achieve the objectives mentioned therein.

2. Trainees must understand the following:—

— the need to be clearly visible to other road users (the use of conspicuity aids);

— the importance of knowing the legal requirements for riding on the road;

— why motor cyclists are more vulnerable than most road users;

— the need to drive at the correct speed according to road and traffic conditions;

— the importance of knowing the Highway Code;

— the need to ride defensively and anticipate the actions of other road users;

— the need to use rear observation at appropriate times;

— the need to assume the correct road position when riding;

— the need to leave sufficient space when following another vehicle;

— the need to pay due regard to the effect of varying weather conditions when riding.

3. Trainees must be aware of:—

— the effect on a vehicle of the various types of road surface that can be encountered;

— the dangers of drug and alcohol use;

— the consequences of aggressive attitudes when riding;

— the importance of hazard perception.

(E) Practical on road riding

1. Trainees must undertake on road riding for a period of not less than two hours. They must (subject to paragraph 2 below) encounter all the following traffic situations and demonstrate their ability to handle each one competently and safely:—

— roundabouts

— junctions

— pedestrian crossings

— traffic lights

— gradients

— bends

— obstructions.

2. Upon application being made by an approved training body for the purpose, the Secretary of State may excuse that body from compliance with a requirement mentioned in paragraph 1 above in respect of practical on-road instruction conducted from premises where the training body provides courses if, having regard to the location of those premises, he is satisfied that it is impractical to comply with that requirement.

3. Trainees must also repeat the following exercises in normal road conditions:—

— carry out a U-turn manoeuvre satisfactorily;

— bring the machine to a stop under full control as in an emergency.

SCHEDULE 13   Regulations 60, 65 and 68

APPROVED MOTOR BICYCLE TRAINING COURSES: FORMS OF CERTIFICATE

PART 1

CERTIFIED INSTRUCTOR'S CERTIFICATE

Road Traffic Act 1988

Certificate No.

Certified Motorcycle Instructor

Certificate of Authorisation

[Photograph of Certificate holder]

Name of Certificate holder...

Name and address of training establishment for which certificate valid

....

Date of expiry

....

The certificate shall indicate, if appropriate, that the holder has successfully completed the Secretary of State's assessment course for certified instructors.

PART 2

CERTIFIED DIRECT ACCESS INSTRUCTOR'S CERTIFICATE

Road Traffic Act 1988

[Photograph of Certificate holder]

Certificate No.

Certified Motorcycle Instructor

Certificate of Authorisation

Name of Certificate holder...

DIRECT ACCESS QUALIFIED

Name and address of training establishment for which certificate valid

....

Date of expiry

....

PART 3

CERTIFICATE OF COMPLETION OF APPROVED TRAINING COURSE

Road Traffic Act 1988

Certificate of Completion of an Approved Training Course for Motor Bicycles

Driver Number of Candidate ....

Date and time of course completion...

Current name ....

Current address..

....

.... [Postcode]..

has successfully completed an approved training course for riders of motor bicycles prescribed for the purpose of Section 97 of the Road Traffic Act 1988 (as amended by Section 6 of the Road Traffic (Driver Licensing and Information Systems) Act 1989).

Signature of certified instructor..

EXPLANATORY NOTE

(This note is not part of the Regulations)

These Regulations consolidate with amendments the Motor Vehicles (Driving Licences) Regulations 1996 and regulations amending those Regulations (all of which are now revoked). The amendments made are as follows:—

    (a) No licence may be issued for sub-category B1 (invalid carriages) to a person who did not hold one on 12th November 1999—although existing licences remain valid—and driving tests for that sub-category are discontinued with effect from the same date (*regulation 5(3) and Schedules 2 and 5*).

    (b) Members of the armed forces are authorised to drive dual purpose vehicles according to the class of licence held and the maximum authorised mass of the vehicle (*regulation 8*).

    (c) A person who has held a licence for category C, D, C + E or D + E for not less than three years in all is qualified to supervise a provisional licence holder in the same category notwithstanding that he has not held a full licence continuously since 6th April 1998 (*regulation 17(3)*).

    (d) The option for a theory test candidate to pay a lower test fee in return for a delay in notification of the result will not be available from 4th January 2000 but from that date all tests must be marked and the results provided on the day of the test. The current lower fee (£15.50) will from that date be payable in the case of every theory test (*regulations 26(1), 27(1), 28(1), 30(1) and 47(1) and (2)*).

    (e) Persons holding a full licence authorising the driving of vehicles in sub-category B1 (invalid carriages) are exempted from the fee when taking a test for a licence authorising the driving of vehicles in category B (*regulation 35(5)*).

    (f) Certain spent provisions are omitted.

In addition, some provisions in Parts II, III and V have been rearranged and some minor drafting changes have been made.

£8.80

© Crown copyright 1999

Printed in the UK by The Stationery Office Limited
under the authority and superintendence of Carol Tullo, Controller of
Her Majesty's Stationery Office and Queen's Printer of Acts of Parliament.
WO 5595   10/99   454591   19585